U.S.S. ENTERPRISE™

2151 onwards (NX-01, NCC-1701, NCC-1701-A to NCC-1701-E)

Owners' Workshop Manual

Ben Robinson and **Marcus Riley**
Technical Consultant **Michael Okuda**

STAR TREK™
OFFICIAL LICENSED PRODUCT

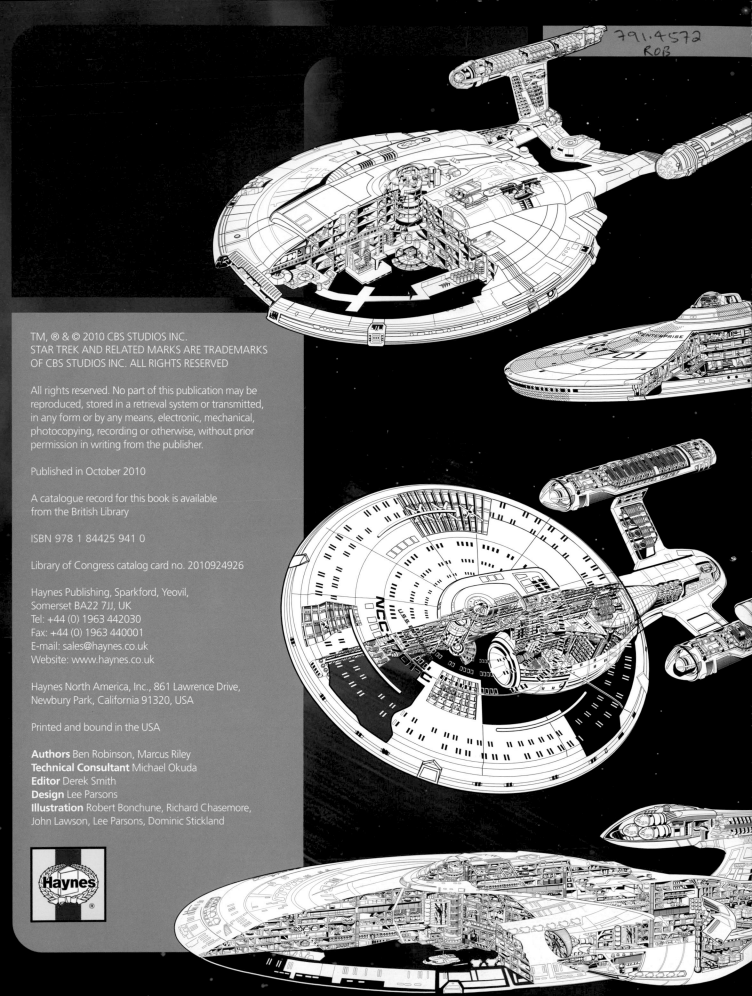

TM, ® & © 2010 CBS STUDIOS INC.
STAR TREK AND RELATED MARKS ARE TRADEMARKS
OF CBS STUDIOS INC. ALL RIGHTS RESERVED

Published in October 2010

A catalogue record for this book is available
from the British Library

ISBN 978 1 84425 941 0

Library of Congress catalog card no. 2010924926

Haynes Publishing, Sparkford, Yeovil,
Somerset BA22 7JJ, UK
Tel: +44 (0) 1963 442030
Fax: +44 (0) 1963 440001
E-mail: sales@haynes.co.uk
Website: www.haynes.co.uk

Haynes North America, Inc., 861 Lawrence Drive,
Newbury Park, California 91320, USA

Printed and bound in the USA

Authors Ben Robinson, Marcus Riley
Technical Consultant Michael Okuda
Editor Derek Smith
Design Lee Parsons
Illustration Robert Bonchune, Richard Chasemore,
John Lawson, Lee Parsons, Dominic Stickland

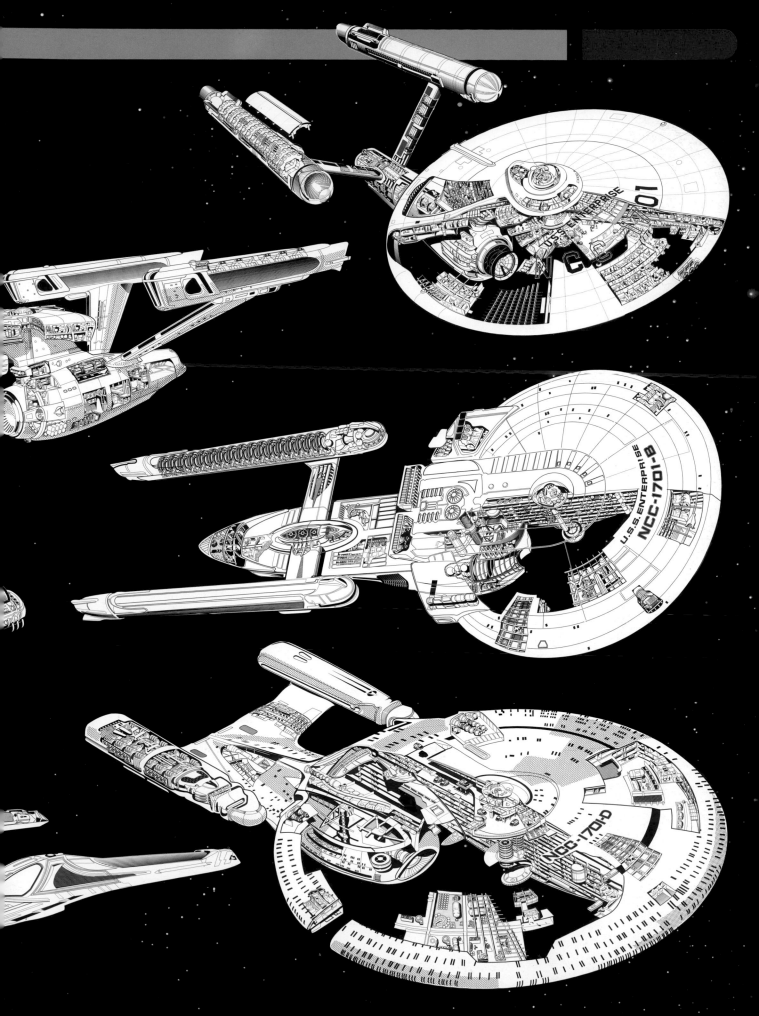

Contents

CONTENTS

Foreword
By Michael and Denise Okuda

One of the best parts about working on the *Star Trek* television shows and movies was that we were occasionally able to sneak a friend onto the Paramount lot to see the "real" *Enterprise*. We weren't supposed to do this, of course, but those sets were *really* cool. Fortunately, our producers usually smiled tolerantly and looked the other way. (Well, at least as long as we didn't do something dumb like walking in while the cameras were rolling!)

As with most soundstages, the *Star Trek* interiors were usually surrounded by a tangle of lights, cables, and various arcane tools of filmmaking, a constant reminder that this was make-believe. Not to mention other small details, like the fact that the sets were obviously made from plywood and that they were nailed to the floor of a Hollywood soundstage.

Nevertheless, there was something very powerful in actually *being there* at Paramount. You could see it in our visitors' eyes as they walked through the corridors of a starship, stood in the transporter room, or even sat in the captain's chair. They'd smile inwardly as they imagined themselves not on a soundstage in Los Angeles, but on a futuristic starship, discovering new worlds in the distant reaches of the galaxy. Whenever that happened, we'd smile because we've been there too. Sure, they knew it was just plywood and plastic and paint, but there was something compelling about those starship interiors that made you want to believe they were real.

A large part of that compelling credibility came from the genius of *Star Trek*'s art director, Walter M. "Matt" Jefferies, the man who designed the original *Enterprise*. The scientific theories behind the show may not have been entirely clear to television audiences in the 1960s, but Matt's background in real-world aviation helped make the ship look believable by infusing his work with what he called "aircraft logic." Matt's innovative yet practical artistry helped to ensure that his ship—and its descendants—would become a cornerstone of *Star Trek*'s magic.

With this book, writers Ben Robinson and Marcus Riley and their team of illustrators have worked hard to capture a bit of that magic so they can share it with a much wider circle of friends. You see, what you hold in your hands is more than just a technical manual. Of course, they've filled this volume with technical data, cool drawings, and specifications of each *Starship Enterprise*. They've also put those ships into context by intertwining "historical facts" from many of our favorite episodes and movies. And that's where it starts to get really fun, because those adventures are what the *Enterprise* is all about.

Let's face it. We love the good ship *Enterprise*, in all her beautiful incarnations. We love her elegant design, we love her technology, and we love the bold adventures that we've shared with the heroes who have voyaged with her. That's why we all want to share Gene Roddenberrry's creation, whether through a furtive studio tour, a series of television episodes and movies, or even through this volume.

Gene was always very proud of his ship and of the fact that it became an international icon of humanity's quest to explore the unknown. But he was even prouder of the effect that his ship had on those who shared his vision of a better tomorrow. *Star Trek* and the *Enterprise* have helped inspire young people around the world to study, to create, to share, and to explore (boldly, of course!). Those young people have grown up to become teachers, doctors, nurses, military officers, artists, scientists, engineers, and even astronauts.

But they all have one thing in common with you and me and with those friends we snuck onto the set. We *all* love the *Enterprise*.

Michael Okuda

Denise Okuda

Los Angeles, California
September, 2010

U.S.S. Enterprise
HISTORY: NX-01—NCC-1701-E

There have been *Enterprises* throughout Earth's history. The French, English, and American navies all had *Enterprises*, from early sloops and schooners to the world's first nuclear-powered aircraft carrier. The name was, however, to become most famous because of the various starships that carried it.

The first *Enterprise* designed for space travel was NASA's prototype space shuttle OV-101. She was unveiled in September 1976 and undertook vital tests, including unpowered landings and stress testing. However, because she was a prototype, this *Enterprise* never left Earth's atmosphere. The honor of being the first *Enterprise* in space would go to the *NX*-class *Enterprise* NX-01, which launched in 2151 under the command of Captain Jonathan Archer. This ship was one of the most important vessels in Earth's history since she was the first starship to be fitted with a Warp 5 engine.

Early Earth ships were capable only of speeds in the region of Warp 2. At this speed a journey between nearby stars could still take several years. At Warp 5 those same journeys would take only a few weeks, so the development of the Warp 5 engine was a huge priority. The Vulcans felt that humanity was not ready to enter the fray of galactic politics and refused to share their superior warp engine designs, so mankind had to develop the technology without their assistance.

Zefram Cochrane, the inventor of the warp engine, broke ground at the Warp 5 Research Complex in 2129, some 66 years after his first historic flight. The most significant work on the Warp 5 engine was done by Henry Archer. His son, Jonathan, a noted test pilot who had been the first man to break the Warp 2 barrier, was assigned as *Enterprise* NX-01's captain. The ship remained under his command throughout her operational lifespan, and her crew made first contact with dozens of intelligent species and famously laid the groundwork for the foundation of the United Federation of Planets in 2161. Miraculously she survived her mission intact and became a museum ship.

The next Starship *Enterprise* was a *Constitution-class* vessel that was launched

2151 NX-01

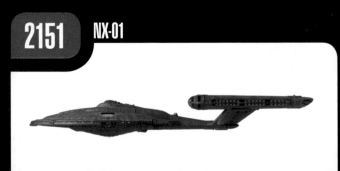

The first Starship *Enterprise* was one of the revolutionary *NX*-class vessels. She was fitted with a Warp 5 engine that enabled her to venture into deep space. Under the command of Captain Archer the ship made first contact with many races and laid the foundations for the United Federation of Planets.

2245 NCC-1701

The second Starship *Enterprise* was a *Constitution*-class vessel that operated in the second half of the 23rd century. During her first 20 years of service she was commanded by Captains April and Pike, but is best known for her service under Captain James T. Kirk. She underwent a major refit in 2270.

in 2245 under the command of Captain Robert April, who completed a five-year mission of deep-space exploration. This *Enterprise* (NCC-1701) was then commanded by two of Starfleet's best-known captains. Christopher Pike assumed command in 2250 and completed two five-year missions that are among the most famous in Federation history. During his command, Pike was joined by the first Vulcan to serve in Starfleet since the foundation of the Federation. The half-human, half-Vulcan Spock was originally the ship's science officer. He would later go on to captain the ship and to become one of the Federation's greatest ambassadors.

At the end of Pike's first mission the ship underwent a major refit that saw the crew complement increase from 203 to 430. After 11 years in command Pike was promoted to Fleet Captain and command passed to James T. Kirk, who became one of the most admired captains in Starfleet history. Kirk's initial five-year mission is required reading at Starfleet Academy and was so successful that he was promoted to Admiral when he completed it in 2269.

The *Enterprise* then underwent an 18-month refit. This was an extensive rebuild and was supervised by the *Enterprise*'s legendary Chief Engineer Montgomery Scott under the command of Captain Willard Decker. When the incredibly powerful entity known as V'Ger approached Earth in 2271, destroying everything in its path, the newly refitted *Enterprise* NCC-1701 was rushed into service, with Admiral Kirk assuming temporary command. The mission was a success, but Captain Decker was lost when he merged

with V'Ger, evolving into a new life form and saving Earth.

In 2277 Starfleet recognized *Enterprise*'s unique contribution to history by abolishing the separate emblems that had been used on different Starfleet ships and starbases and replacing them with the *Enterprise*'s 'arrowhead' badge.

By 2284 *Enterprise* had been assigned to Starfleet Academy under the command of Captain Spock as a training vessel. The following year, she responded to a signal from the space station Regula I, where Dr Carol

The first Starship *Enterprise* was an *NX*-class vessel that was fitted with a breakthrough Warp 5 engine. It made 'Enterprise' one of the most famous names in Starfleet's history and set a standard that its successors would have to emulate.

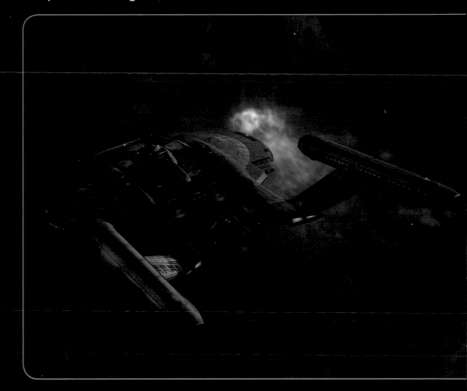

2271 — NCC-1701-REFIT / A

The NCC-1701 (refit) was rushed into service in 2271 to confront the V'Ger entity. After her destruction in orbit around the Genesis planet, Starfleet renamed the U.S.S. *Yorktown* as the *U.S.S. Enterprise* NCC-1701-A in her honor. Launched in 2286, the 1701-A had undergone a major refit that saw all her systems significantly upgraded.

2293 — NCC-1701-B

The *U.S.S. Enterprise* NCC-1701-B was an *Excelsior*-class ship that was nearly lost before she entered active service: during the dedication ceremony, the ship responded to a distress call and Captain Kirk, who was a guest of honour, was lost saving two El-Aurian ships. The ship survived and remained in service until 2329.

Marcus was developing the Genesis device. This was a terraforming technology that could completely restructure a planet's environment, making a lifeless world habitable. Unfortunately Genesis had the side effect of destroying any life that had existed on the planet before it was deployed. When *Enterprise* responded to the signal, Spock passed command to Admiral Kirk, who was on board, arguing that it was only logical.

During the mission the Genesis device fell into the hands of Khan Noonien Singh, a genetically engineered madman from Earth's past. Kirk prevented him from using Genesis, instead detonating it in the Mutara Nebula, where it created a new planet. Unfortunately, Captain Spock was killed during the mission and his body was laid to rest on the new Genesis planet.

Kirk returned the badly damaged *Enterprise* to Spacedock, where Starfleet decided that the damage was so severe that the ship should be retired. However, when Kirk learned that Vulcan tradition required him to return Spock's body to Vulcan he and several members of his senior staff disobeyed direct orders, stole the *Enterprise*, and returned to Genesis. When they arrived, Kirk was forced to destroy the *Enterprise* to prevent the Klingons taking control of her.

Incredibly the Genesis effect had regenerated Spock's body and the priestess T'Lar used the ancient *fal-tor-pann* ceremony to rejoin it with his *katra* on Mount Selaya on Vulcan. In 2286 Kirk and his crew returned to Earth to face court martial, but when they arrived the planet was under attack by an alien probe and they were instrumental in saving it.

In recognition of their extraordinary service, Starfleet assigned them to another Constitution-class ship, the *U.S.S. Yorktown*, which was renamed the *U.S.S. Enterprise* NCC-1701-A in honor of her predecessor. Kirk was demoted to captain for disobeying orders and assumed command of the ship for the rest of her service until she was retired in 2293.

A new *Excelsior*-class *Enterprise*, with the registry number NCC-1701-B, was launched the same year. Kirk attended the dedication ceremony but was lost, presumed killed, when *Enterprise* rescued two El-Aurian vessels from an energy disturbance. This *Enterprise* served with distinction, her crew boldly exploring the unknown area of space beyond the Gourami Sector, and she was one of the several vessels involved in the Tomed Incident, which led to the re-establishment of the Romulan Neutral Zone in 2311. She was finally lost in action in 2329. Her fate is uncertain but it is assumed that the crew contracted a plague.

The next *Enterprise* (NCC-1701-C) was an *Ambassador*-class ship that was launched in 2332 under the command of Captain Rachel Garrett. This ship was destroyed in 2344 defending a Klingon outpost at Narendra III from a Romulan attack. The loss of *Enterprise* NCC-1701-C played a vital role in establishing peace between the Federation and the Klingon Empire, since the Klingons greatly admired the crew's willingness to sacrifice their ship in the face of insurmountable odds in an effort to save a small number of Klingons.

Starfleet took the decision to reserve the name *Enterprise* for one of the *Galaxy*-class ships which were then in development at the Utopia Planitia shipyards above Mars. Work on the *Enterprise* NCC-1701-D, the third *Galaxy*-class ship, began in 2350. The new *Enterprise*

2332 — NCC-1701-C

The *Enterprise*-C played an important role in cementing relations between the Federation and the Klingon Empire. Under the command of Captain Rachel Garrett she defended a Klingon outpost from a Romulan attack. The ship was destroyed in the battle but the crew's sacrifice was admired by the Klingons.

2363 — NCC-1701-D

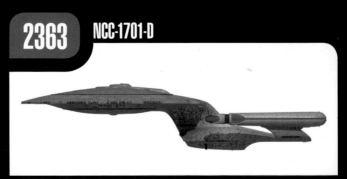

The *Enterprise*-D was a radical departure from its predecessors since this ship had a crew of over a thousand including a large civilian population. She was commanded by Jean-Luc Picard and was Starfleet's flagship. She was lost in 2371 when she was attacked by the Duras sisters.

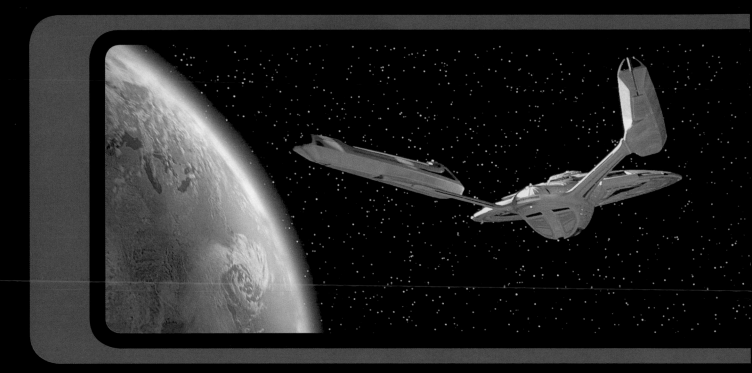

wouldn't launch until 2363, but she was made Starfleet's flagship and placed under the command of veteran Captain Jean-Luc Picard, one of the most admired officers in Starfleet.

The new *Enterprise* was a significant departure from her predecessors since she had a crew of over a thousand, which included a large civilian population and their families. The *Enterprise*-D lived up to her famous name and was instrumental in saving the entire Alpha Quadrant from the Borg invasion of 2366 as well as playing a vital role in the Klingon Civil War of 2368.

The ship was destroyed in 2371 when a renegade Klingon faction attacked her, causing her warp core to overload. The crew survived after separating the saucer section and landing it on the planet Veridian III.

Starfleet immediately commissioned a new *Enterprise*, the *Sovereign*-class *U.S.S. Enterprise* NCC-1701-E. Again, Picard was given command, and he was joined by a significant proportion of his original crew, including all his senior staff. This *Enterprise* was involved in saving Earth from attempted invasions by the Borg and the Romulans. In 2379 she was due to explore the Denab system, following in the footsteps of her distinguished predecessors.

By the late 24th century seven different starships had carried the name 'Enterprise'. The *Enterprise*-E has much the same mission goals as the original NX-01 and is Starfleet's flagship.

2372 NCC-1701-E

The *Sovereign*-class *Enterprise*-E was originally designed to fight the Borg and was a leaner ship than her predecessor. She launched in 2372 with the same senior staff as the *Enterprise*-D. A year later, she helped save Earth from a second Borg invasion and then in 2379 prevented a Romulan attack.

Enterprise
NX-01

The *Enterprise NX-01* was one of the most significant ships in the history of Starfleet and played a vital role in the establishment of the United Federation of Planets. She launched in April 2151 and remained in service for ten years before being mothballed after advances in warp technology rendered her revolutionary engine obsolete. During those ten years *Enterprise NX-01* was involved in dozens of first-contact missions, saved Earth from destruction at the hands of the Xindi, altered the course of Vulcan politics, and played a pivotal role in establishing treaties with the Andorians and Tellarites. The NX-01 went on to become a museum ship that is still being visited hundreds of years later.

Classification	NX class
Constructed	Earth Spacedock
Launch date	2151
Retired	2161 [became a museum ship]
Length	225m
Number of decks	7
Crew complement	83
Weaponry	Phase cannons and nuclear torpedoes [2151]; photonic torpedoes [2153]
Commanding Officers	Jonathan Archer

The *Enterprise* NX-01 was the first ship in the NX class, which proved so important to Earth's history. These revolutionary vessels were humanity's first spaceships to be fitted with the Warp 5 engine that made interstellar exploration a practical reality. *Enterprise* NX-01 was rushed into service three weeks before she was ready so that she could return Klaang, a Klingon warrior who had crashed on Earth, to his people. As a result she hadn't completed her shakedown cruise, but since the Klingon mission was deemed a success Starfleet ordered *Enterprise* NX-01 to commence her mission.

Enterprise NX-01 was commanded by Captain Jonathan Archer throughout her ten years of active service. He had assembled the finest crew available to Starfleet and they all became almost legendary figures in the history of space exploration. The ship's chief engineer was Commander Charles 'Trip' Tucker III, a long-time friend of Archer's, who made major alterations to the Warp 5 engine during his ten years onboard and had a profound influence on the design of all subsequent Starfleet vessels. The armory officer, Lt Malcolm Reed, came from a naval

family; during the mission he helped Tucker make significant improvements to the design of phase cannons, and developed Starfleet's first effective force fields. The communications officer, Hoshi Sato, was a linguistic prodigy who grasped new languages with incredible speed and played a vital role in expanding the data set that the Universal Translator used. When *Enterprise* NX-01 launched, her pilot and navigation officer Travis Mayweather had more experience of deep space than any other crewmember, since he had been born and raised on a cargo freighter. The senior staff was rounded out by Dr Phlox, a noted research scientist from Denobula, and T'Pol, a Vulcan who was originally assigned by the Vulcan High Command to advise *Enterprise* NX-01's human crew, although she subsequently left the High Command and joined Starfleet.

Enterprise NX-01 rapidly earned her place in history. The plan was for her to go to Warp 5 shortly after leaving Earth's solar system, but in fact she wouldn't achieve this historic feat until February 2152. However, her regular cruising

The *Enterprise* NX-01 was constructed in orbit around Earth and launched in 2151. She was a revolutionary ship, fitted with a functioning Warp 5 engine, and ushered in a new era of deep-space exploration.

speed was in excess of Warp 4, which was sufficient to reach planets in nearby star systems in just a few weeks rather than months or years.

The crew found their first Minshara-class planet (*ie* a planet suited to humanoid life) three weeks after leaving Earth, would go on to visit dozens of inhabitable worlds, and Archer would go down in history as one of the greatest explorers of the 22nd century. Two planets and countless schools and other institutions on Earth would be named after him.

Enterprise NX-01 also played a significant role in Earth–Vulcan relations. At this period in history Vulcan and Andoria were on the verge of open war. Despite the Vulcans' public protestations of innocence, Archer exposed a Vulcan spy station in the ancient monastery at P'Jem that was aimed at Andorian space. This earned Archer the trust of an Andorian captain named Shran. Over the following years Archer was able to establish Starfleet as an impartial intermediary between the Vulcans and the Andorians and later between the Andorians and the Tellarites. This laid important groundwork for the creation of the United Federation of Planets.

Archer's dealings with other civilizations were not all as successful. As has been mentioned, the *Enterprise* NX-01's first mission was to return the stranded Klingon warrior Klaang to his home world, a gesture that proved significant in helping to prevent Earth's contentious relationship with the Klingon Empire from descending into open warfare. However, the ship's subsequent encounters with the Klingons were not particularly successful and Archer was unable to establish good diplomatic relations with them. Over the next century the Klingons would become one of the greatest threats to the Federation.

Enterprise NX-01 was also involved in countering one of the most significant military threats of the 22nd century. In March 2153, the Xindi launched a devastating attack on Earth. A Xindi weapon entered Earth's orbit and fired a force beam that cut a 4,000km-long swath from Florida to Venezuela, killing seven million people. Earth soon learned that this weapon was only a prototype and that the Xindi were building a larger version that would destroy the entire planet.

The NX-01 was recalled to Earth and fitted with additional weaponry before being sent to find the Xindi and stop them. Archer discovered that the Xindi were being manipulated by creatures from a different dimension and—according to some reports—from a different time. He managed to persuade a faction of the Xindi to side with him and prevented the weapon from being fired, saving Earth from destruction. This ensured that the *Enterprise* NX-01 and her crew would never be forgotten.

In 2154 *Enterprise* NX-01, and Commander T'Pol

in particular, became involved in the overthrow of the Vulcan High Command, which had become increasingly aggressive and was suppressing the Syrannite movement, which believed that Vulcan had moved away from the teachings of Surak, a figure from Vulcan history who had persuaded them to abandon violence and dedicate themselves to logic. Captain Archer was instrumental in recovering Surak's original writings, the *Kir'Shara*, and proved that V'las, the leader of the Vulcan High Command, had fabricated evidence to justify a pre-emptive strike on Andoria. V'las was replaced by Kuvak, who dissolved the High Command and pursued a much more positive policy toward Earth.

Despite the dangerous nature of space exploration, Archer and his senior staff almost all survived the ten years of their mission. The one significant exception was Commander Tucker, who was killed in 2161 on the ship's last mission, saving the Andorian Shran from a hostile boarding party. *Enterprise* NX-01 returned to Earth in time for Captain Archer to sign the charter that founded the United Federation of Planets. Archer would later be promoted to Admiral and become Chief of Staff at Starfleet Command. In 2164 he was made an honorary member of the Andorian Imperial Guard and served for five years as Federation Ambassador to Andoria. He would eventually become a member of the Federation Council, serving as its president for eight years from 2184 to 2192.

However, despite all his later achievements he will always be remembered as the first captain of the Starship *Enterprise*.

The *Enterprise* NX-01's original mission was one of peaceful exploration but she soon found herself in conflict with numerous races and was often attacked. As a result Starfleet upgraded the ship to give her greater combat abilities.

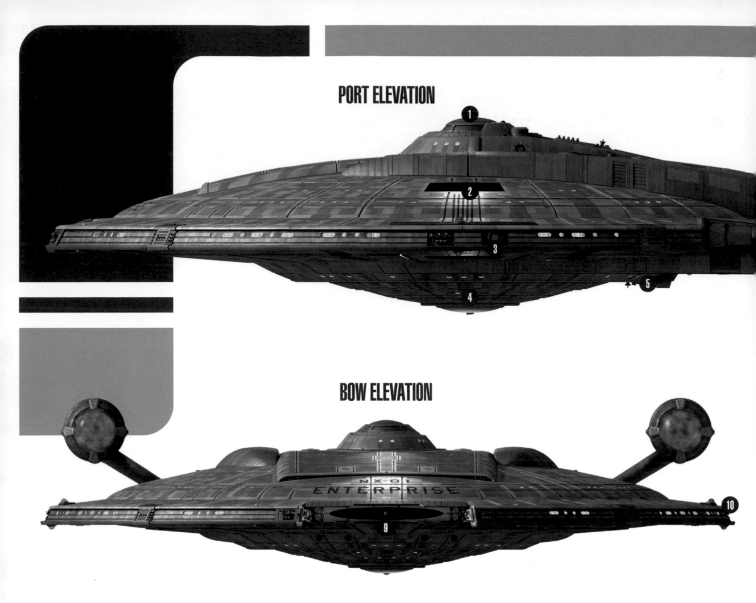

PORT ELEVATION

BOW ELEVATION

NX-01
ENTERPRISE

The *Enterprise* NX-01 was the prototype *NX-class* vessel. She was 225 meters long and seven decks high; when she launched she had a crew of 83, though this was later supplemented with a MACO (Military Assault Commando) unit. When she launched in April 2151 she was by far the most advanced Earth vessel ever built. The warp engines were capable of Warp 5, which made interstellar exploration a practical reality. She was equipped with state-of-the-art sensors, a computer core than ran through three decks, and one of the first transporters passed for bio transport. She was also designed to carry experimental phase weaponry, though this wasn't fitted for several months.

The *NX* class established the configuration that would become a trademark of Starfleet vessels, with a forward section that housed the crew quarters and the majority of the ship's habitable volume. The rear section contained the engineering systems and was connected to twin

warp nacelles, which were supported on pylons to keep them away from the main body of the ship.

When in dry dock the entire rear hull section could be removed and the warp core and the engineering-room modules could be replaced. Behind the hull, between the warp nacelles, there was a large Symmetrical Warp Governor. This was used to regulate the warp fields generated by the nacelles, which were unequal. It did this by generating a low-yield subspace field which 'bent' the warp bubble into a more acceptable configuration for high warp velocities.

The seven decks were labeled A through G with the bridge on the uppermost level on A Deck and the ventral sensor array on G Deck. Two half-decks were located between D and E Decks and between E and F Decks. These contained a combination of EPS conduits, air ducts, and access tubes.

The engine (more correctly referred to as the gravimetric field displacement manifold) was based around a primary warp coil with five plasma

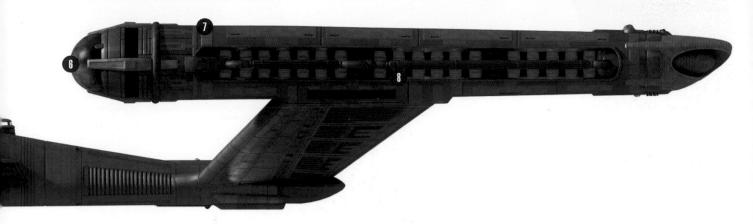

STERN ELEVATION

injectors. It was housed in main engineering just behind the centre of the saucer section and crossed C and D Decks.

It worked on the same principles as the original Cochrane engine by generating a matter/ antimatter reaction. That reaction was focused through dilithium crystals and the positron stream was aligned with a series of magnetic constrictors. The matter used was deuterium (a hydrogen isotope with a single proton and a single neutron in the nucleus with a single electron). The main reactor contained enough matter and antimatter to run for approximately a month without being resupplied. *Enterprise* NX-01 also carried over 2,000 liters of deuterium, which was enough to keep the engine operating for extended periods away from Starfleet facilities. The deuterium storage tanks were on B Deck in the raised fairings that are visible from the outside of the ship. The NX-01 carried a similar amount of antimatter, which was stored in pods on F Deck.

1. **Main bridge**
2. **Cargo-bay doors**
3. **Airlock**
4. **Sensor dome**
5. **Shuttlebay**
6. **Bussard collector**
7. **Warp nacelle**
8. **Warp coils**
9. **Navigational deflector**
10. **Hard connect point**
11. **Impulse engine**

In case of emergencies the warp engine could be shut down manually by controls on the top of the main reactor. This disengaged the plasma injectors, causing the warp reaction to collapse.

When the engine was fired up it took 20 minutes for the nacelles to reach full power, although the ship could go to warp considerably sooner than this.

When *Enterprise* NX-01 was launched she experienced noticeable fluctuations in the warp field when exceeding speeds of Warp 4.8. These caused the ship to shake, producing an uncomfortable ride and the sensation that the ship was pulling herself apart. In early 2153 Commander Tucker discovered that he could eliminate these fluctuations by compressing the antimatter stream before it reached the injectors, creating a much more stable warp field. Later the same year he was able to upgrade the antimatter injectors, using a design given to him by the Andorians that featured variable compression nozzles.

Superheated plasma was channeled from the engine into the plasma accelerators in the raised double hulls on the saucer section. From here it was fed through plasma conduits in the struts that connected the saucer to the engineering section, and on to the nacelles. The nacelles were not normally

1 Ship's registry

2 Cargo-bay doors

3 Impulse engines

4 Warp nacelle

5 Symmetrical warp governor

6 Observation gallery

7 Shuttlepod launch bay

8 Nacelle pylon

DORSAL ELEVATION

SYSTEMS OVERVIEW

The NX-01's warp engine was a major advance but it was also an experimental piece of technology that required a substantial amount of maintenance and regular attention. As a result, working in engineering was not always safe.

VENTRAL ELEVATION

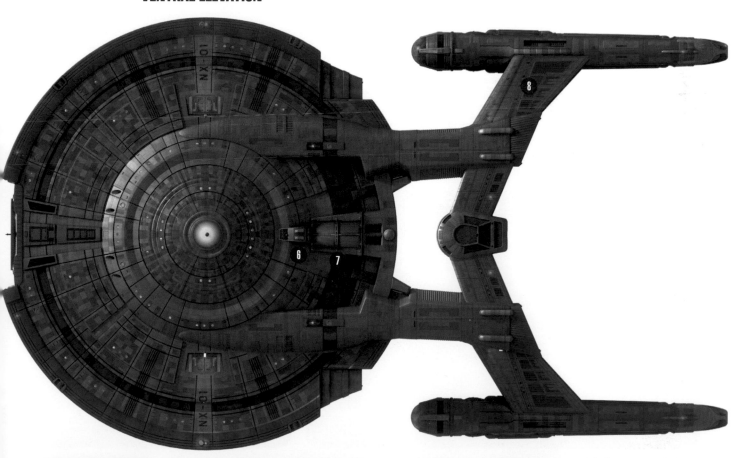

accessible to the crew when the engines were running, since the plasma reached temperatures in excess of 15 million kelvin, but a catwalk ran the length of each nacelle. The nacelle enclosures were lined with 25-micron osmium diffraction foil, which served to contain charged particle emissions when the warp propulsion system was powered down or when containment flux was below operational levels.

An auxiliary power system was completely independent of the main grid and functioned without any power from the warp engine. Like the main power supply, this supplied the EPS (electro-plasma system), which distributed power around the ship.

The *Enterprise* NX-01's outer hull was constructed of tessellated tritanium alloy plates and was lined with duranium. The *NX* class used polarization rather than shields to strengthen the hull when the ship was attacked. The polarization worked by sending an electromagnetic charge through the hull plating to make it significantly stronger and more resistant to impact or directed energy weapons. The electromagnetic charge had to be maintained or the hull would return to normal strength.

During her ten years in service, *Enterprise* NX-01 endured many hull breaches. In the event of such serious damage the affected area was sealed by emergency bulkheads that isolated the depressurized area. The ship carried spare hull plating and had the facility to fabricate more, but on several occasions she sustained serious damage that forced her to return to Spacedock.

In the event of a catastrophic emergency the crew could abandon ship using escape pods located around the rim of D Deck. Each pod was designed to carry two people and could separate from the ship at a relative speed of 300kph. They were designed to float in space emitting an emergency signal that could be picked up by nearby vessels.

Enterprise NX-01 carried two flight recorders, or black boxes, that constantly recorded data from the captain's log, the ship's sensors, and the main engineering systems. They were designed to withstand a warp engine breach and emitted a locator signal with a range of five million kilometers.

The NX-01's primary mission was deep-space exploration and the ship's sensors were the most sophisticated that Earth had to offer, although they were not on a par with contemporary Vulcan or Andorian systems. They could scan the full electromagnetic spectrum and could measure subspace and gravimetric field gradients. Sensors could also characterize the biosignatures of a wide range of life-forms. The main sensor arrays were located in pallets on the edge of the saucer section, next to the deflector dish, in the massive dome on the underside of the saucer, and in a smaller dome on the top of A Deck just behind and to the port of the bridge.

Enterprise NX-01 was one of the first Starfleet vessels to be fitted with a transporter that had been cleared for use by living beings. It could be found on D Deck. The transporter had a range of approximately 2,000km, but since transporter technology was still in its infancy the crew relied on shuttlepods to travel to the surface of planets and to other vessels. NX-01 was equipped with two shuttlepod launch bays, both located on D and E Decks, near the aft of the saucer module. Two pairs of launch doors on the underside of the saucer hull allowed shuttlepods to drop from the ship and to be recovered after each mission.

The shuttlepods were deployed by means of magnetic clamps that descended from the bay ceiling. During a shuttle launch, the relevant shuttlebay was completely depressurized. Crewmembers could supervise operations from the pressurized observation room inside the shuttlebay.

Below: When she was traveling at sublight speeds, the NX-01 relied on her impulse engines, which provided thrust in the same way that rockets had on earlier craft.

Below right: Attitude and fine positional control, as well as small translational maneuvers, were accomplished through the use of the Reaction Control System (RCS) thruster assemblies.

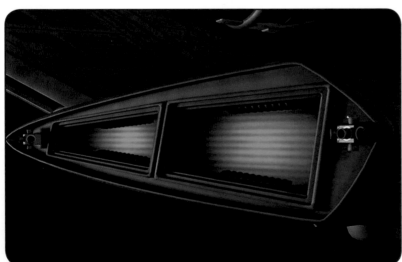

IMPULSE ENGINE TECHNOLOGY

The NX-01 had two impulse engines, which were both located on E Deck. The combined thrust vectors of the two impulse engines normally ran through the ship's center of mass, but could be adjusted slightly to steer the vessel, and could also be adjusted to compensate for non-standard weight distribution.

At the heart of each engine were nuclear fusion reactors that were powered by deuterium. Each engine assembly consisted of three major components: twin fusion reaction chambers, a set of subspace field coils, and an exhaust direction module.

The fusion reaction generated superheated plasma, which was then channeled through subspace compression coils, which increased the apparent mass of the plasma exhaust products, so the resulting forward propulsion of the vessel was as if a much greater mass of plasma had been expelled. This greatly reduced the fuel (and mass) requirements for impulse propulsion.

Finally, the plasma was vented out an exhaust, pushing the ship along normal Newtonian principles, basically like a very powerful rocket. Vectored Exhaust Director Coils were used to alter the direction of the plasma, thus altering the course the ship took.

The impulse engines also provided a secondary power system for the ship, operating as a backup to the warp drive. Plasma was fed from the fusion reactors into the ship's EPS (electro-plasma system) network, meaning that the ship's systems could function normally while the warp engines were offline.

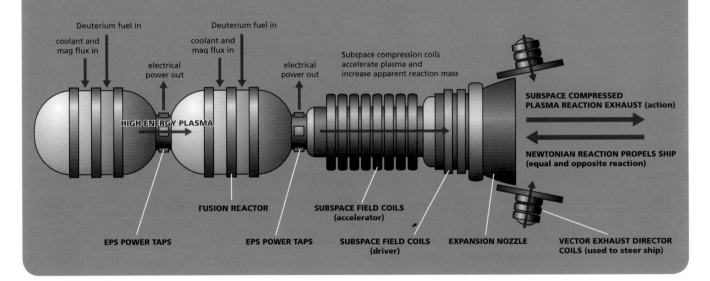

Once a shuttle was launched it could be seen from an observation gallery on F Deck.

Shuttle docking was normally controlled by the shuttle's pilot, but *Enterprise* NX-01 could also retrieve the shuttles by using grapplers that were fired from turrets on the underside of the saucer. The grapplers themselves used powerful electromagnets that could latch on to any ferrous object, and the cables were made of a composite carbon fiber that was strong enough to hold a ship that was moving at speeds approaching full impulse.

Any personnel using the transporter, or the shuttlepods, had to be decontaminated to prevent them from bringing diseases or parasites aboard, so any crewmembers returning to the ship had to spend time in the decontamination chamber, which was also on D Deck. While there they were bombarded with low-level radiation that eliminated alien bacteria or small life forms. Crewmembers also applied decontaminating gels to their bodies that sterilized their skin and killed a wide range of germs.

In addition to the shuttlepods, *Enterprise* NX-01 carried an inspection pod, which was used to assess damage to the hull and carry out repairs to the ship. This was launched from the port cargo bay.

The NX-01 was equipped with conformal airlock interfaces to allow it to dock with a wide variety of vehicles, even those of unknown design. The ship had three main airlocks with docking clamps that would lock two ships together. The most commonly used ones were on opposite sides of E Deck at the outer edge of the saucer, in line with the center of the bridge. Another airlock could be found at the rear of the saucer section.

When the NX-01 was in Spacedock she was held in place by hard-connect points on the outer edges of the saucer immediately above the airlocks.

An important part of *Enterprise* NX-01's mission was establishing a subspace communications network that allowed the ship to maintain contact with Starfleet and improved Earth's ability to stay in touch with distant ships and colonies. She was fitted with a powerful subspace antenna and regularly deployed subspace amplifiers that could relay signals to Starfleet. The subspace amplifiers were stored in bays on F Deck to the port and starboard of the main shuttlepod launch bay and were deployed through hatches on the underside of the saucer.

ENVIRONMENTAL

The environmental systems recycled the air, water, and waste products, with biochemical reactors removing carbon dioxide from the air and producing oxygen in much the way that plants do on Earth. The recycling system was so efficient that water was plentiful and senior staff had showers in their quarters.

CARGO AND STORAGE

Enterprise NX-01 had three cargo bays; after the Xindi attack on Earth, one of the cargo areas was converted into a command centre where the crew could analyze all the data they collected on the Xindi threat. The two main cargo bays ran through the height of the ship from decks C to E, with large cargo-bay doors on the upper and lower surfaces of the saucer. Internally, the cargo bays were accessed from E Deck.

WEAPONS AND DEFENSIVE SYSTEMS

Originally Captain Archer had argued against providing the ship with heavy armaments, but after his first three years in command and his historic mission to find the Xindi, he revised his position, advising that all future *NX*-class vessels should be armed with ventral and dorsal torpedo launchers as well as phase cannons.

Initially the ship was equipped with rocket-propelled torpedoes that used explosive charges. However, because *Enterprise* NX-01 entered service early the torpedoes were not at first usable since

The original plan didn't call for the NX-01 to be heavily armed but her defenses were soon upgraded. Her original primary weapons were torpedoes, which were loaded into the torpedo tubes from the armoury on F Deck.

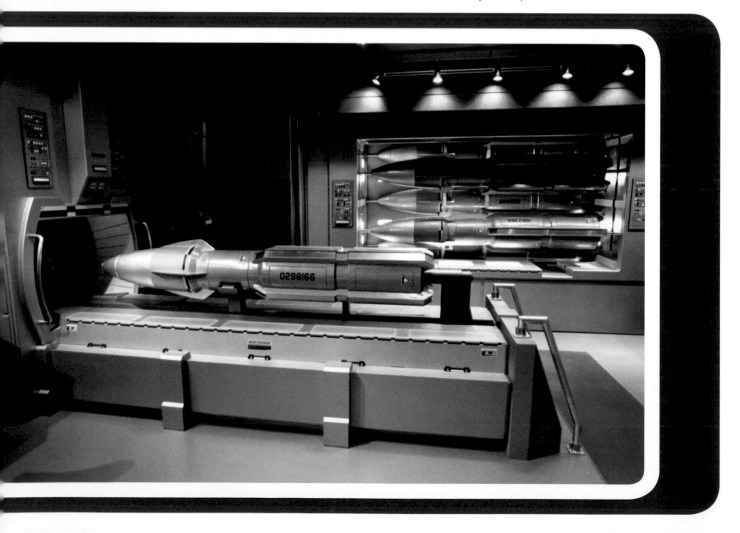

SYSTEMS OVERVIEW

The NX-01 was a relatively small ship with a crew of 83. The ship was nothing like as luxurious as her successors and had a functional design that made limited concessions to the crew's comfort.

the targeting scanners weren't properly aligned, and corrections had to be made after the ship was under way. The torpedoes had an on-board engine and received telemetry from the ship that allowed them to alter course to track a target. They could be remote-detonated.

Enterprise NX-01 was also designed to carry three phase cannons. These were phase-modulated energy weapons with a maximum power output of 500 gigajoules. This technology was a recent development and had only just been certified for use. When NX-01 left Spacedock she had only a single prototype aboard and that wasn't actually fitted.

For the first few months of her mission, NX-01 operated without phase cannons. But after a hostile encounter in August 2151 Captain Archer decided it was time to install them. The original plan was to return to Jupiter Station, but because the ship was in danger the engineering crew began work during the journey. They succeeded in installing the two forward phase cannons on September, 1, just two days after the ship had been attacked.

During the installation, Commander Tucker and Lt Reed calculated that the cannons could take a far greater load than they were designed for if they were tied directly into the ship's impulse engines. The initial test resulted in a plasma recoil in the ship's power systems, but Tucker devised a way to redirect this through the gravity plating, which was then depolarized to absorb the recoil. The excess energy was sent to the structural

integrity field, which had the advantage of making the ship more resistant to damage.

The targeting scanners and control systems for all the weapons were controlled from the ship's armory on F Deck. The torpedoes were stored and loaded from here into the fore and aft tubes. The phase cannons were mounted on the outside of the ship, but their frequencies were set and modulated from the armory.

In addition there were 14 weapons lockers on the ship that were stocked with small arms. A brig on E Deck was designed to hold two people but could accommodate more in an emergency. The walls of the brig were made of reinforced transparent aluminum and were soundproofed. Speakers and microphones built into the walls allowed personnel to communicate with any prisoners.

The NX-01 was fitted with transporters but when she started her mission, shuttles were the principal means used to travel to a planet's surface. They were launched from a drop bay on the underside of the saucer.

MAIN BRIDGE LAYOUT

A Deck was the smallest of the seven decks and consisted of the bridge module and the captain's ready room. It was served by a single turbolift that opened on to the back of the bridge on the port side. An emergency hatch provided access to the rest of the ship if the turbolift failed.

As would become typical on Starfleet vessels, the bridge was circular, with the captain's chair in the center of the room. A single helm station was positioned directly in front of this, between the captain and the main viewscreen, and a variety of stations ran around the outside of the room. Working clockwise from the turbolift, they were the

science station and communications on the port; then on the starboard side there was a tactical station and an engineering station. At the rear of the bridge was a situation room with a tabletop station that was tied into stellar cartography, the ship's library computer, and the engineering systems. The bulkheads in this area incorporated a number of status monitor displays.

The helm station combined navigational and piloting functions. The astrogator was used to enter the ship's heading and to input course corrections. The helm officer also had direct control of the ship's speed, with control of the warp and impulse engines and the thrusters, which were used for close maneuvers. The

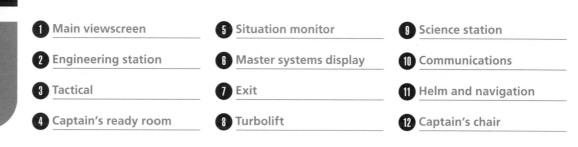

1 Main viewscreen

2 Engineering station

3 Tactical

4 Captain's ready room

5 Situation monitor

6 Master systems display

7 Exit

8 Turbolift

9 Science station

10 Communications

11 Helm and navigation

12 Captain's chair

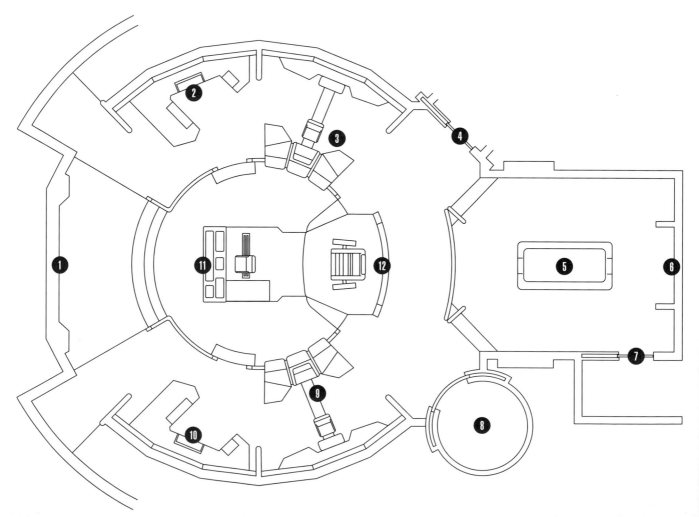

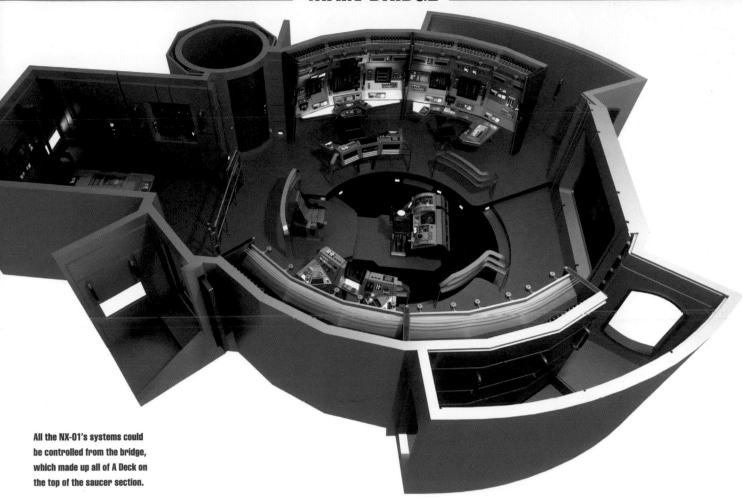

All the NX-01's systems could be controlled from the bridge, which made up all of A Deck on the top of the saucer section.

helm station was in constant communication with main engineering and was provided with a stream of data about the state of the engines, structural integrity, and the navigational deflector.

The science station's principal function was to supply information and analysis from the ship's sensors to the commanding officer and key departments and to compare data with *Enterprise*'s library computer. It provided a massive range of information, including full spectrum readouts from the sensors. The tie-in to the library computer gave it access to all the information in the Starfleet database.

Enterprise's communications systems were the most sophisticated available at the time, but at this point in Starfleet's history communication with other species was not always straightforward. *Enterprise*'s systems were fitted with a Universal Translator, which could analyze the grammar and syntax of any alien language encountered and compare it with other known languages to provide an instantaneous translation. However, when *Enterprise* launched, the UT database contained a relatively small number of languages and it could often only provide very crude translations. Since the crew encountered dozens of new alien races the effectiveness of the UT improved dramatically during the ship's mission.

The tactical station provided direct control of *Enterprise*'s torpedoes and phase cannons. This station was in direct communication with the armory and had full access to the ship's targeting scanners, and to sensors that provided detailed information on nearby vessels, including high-resolution DNA scans of their crews. The targeting scanners were also used to control the ship's grapplers, which in addition to recovering shuttlepods could latch on to any alien vessels or objects that the captain wanted to bring aboard. Captain Archer even employed them as a weapon once by using them to damage the warp nacelle of a Klingon warship.

The tactical station also functioned as a damage control station, with the tactical officer providing the captain with information regarding the status of the hull plating, structural damage to the ship, and casualties.

A doorway on the starboard led to the captain's ready room, which was effectively a small office where he could file reports, receive visitors, talk to staff in private, and relax when he wasn't needed on the bridge. Since it was only meters away from the bridge, the captain could be summoned at literally a moment's notice.

The crew was quartered in the saucer section. Approximately one-third of the crew was female, and two members were non-human—the Vulcan Science Officer T'Pol and the Denobulan doctor Phlox.

The majority of the living quarters was on D Deck, though crewmembers were also billeted on B, C, and E Decks, and there were guest quarters—used by visiting delegations—on G Deck, next to the sensor dome. Commander Tucker and Lt Reed both had their quarters on B Deck, while Captain Archer had picked out quarters on E Deck to the starboard of the main deflector that gave him an exceptional view of the activity around the ship.

Senior officers had individual quarters that were divided into two distinct areas, and had their own bathrooms and showers. More junior crewmembers shared quarters, with two people sharing one set of rooms in which bunk beds were fitted into the bulkhead.

Food was prepared in a galley on E Deck by the ship's chef, who Captain Archer personally selected. The crew met to eat in a single mess hall that also doubled as a recreation area, where they watched movies once a week. Captain Archer also had a private dining room where he would eat with members of the senior staff or with visiting aliens. The ship carried a supply of real food, which was kept in cryogenic storage, but most of the foodstuff was in the form of resequenced protein that could be made to approximate almost any dish. *Enterprise* NX-01 carried enough supplies to last for several years. Other recreational facilities on the ship included a gym on C Deck.

Sickbay was located in the safest part of the ship in the centre of E Deck. It also served as a medical research facility and was provided with state-of-the-art equipment under the supervision of Dr Phlox, who supplemented conventional medical supplies with a menagerie of alien creatures that had medical properties. In the centre of the circular room was an examination bed, which could be sent into the imaging chamber in the bulkhead. This provided the most detailed medical analysis available at the time, with high-resolution scans

Captain Archer personally chose his quarters on E Deck because they provided his favourite view. He also had a ready room, or office, off the bridge, where he performed most of his administrative duties.

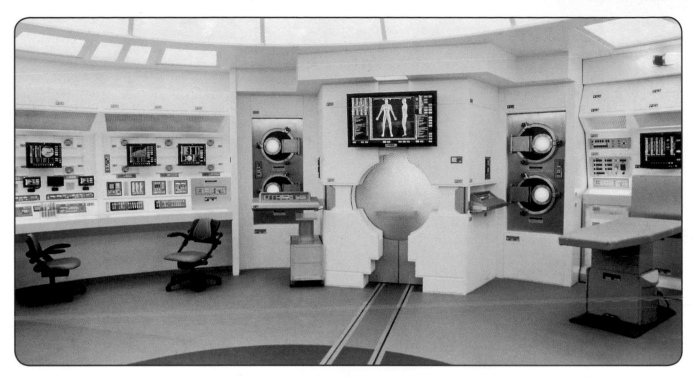

showing the condition of the body down to submolecular level and full DNA analysis. Because of its high resolution, the chamber was also used to perform autopsies.

Three biobeds were positioned next to one another around the edge of the room. This number would later prove inadequate, but when the NX-01 was launched Starfleet did not anticipate that she would be involved in full-scale combat. The Chief Medical Officer could access the environmental systems from sickbay if, for example, he needed to introduce compounds into the ship's atmosphere, but this involved re-routing access from the bridge.

The NX-01's sickbay was state of the art when the ship was launched in 2151 and was fitted a wide array of diagnostic and scanning equipment.

The sickbay was also one of the NX-01's major scientific labs and the ship's science officer T'Pol and chief engineer 'Trip' Tucker often joined Dr Phlox to analyze data or adjust the machinery.

The two shuttlepods were designed to carry a pilot and up to six passengers. They were rated for one-quarter impulse and were used for both ship-to-ship and ship-to-planet transport. They were highly maneuverable and could not only land on a planet's surface but could also enter the tail of a comet or navigate through an asteroid field. Their hulls could be polarized like that of the *Enterprise* NX-01 herself, in order to endure the stresses of entering a planet's atmosphere.

The engine pod was a detachable module that used nuclear fusion to produce thrust. The life-support system relied on an entirely independent power supply so that it could continue operating in the event of a complete engine failure.

The main hatch was on the port side of the shuttlepod, and an extensible airlock was accessed through a hatch in the roof. Small docking clamps on the shuttlepod's exterior allowed it to connect to almost any vehicle.

Shuttlepods were not intended for extended use—they did not have full toilet facilities—but the on-board air tanks and atmospheric recyclers could maintain a breathable atmosphere for several days. The shuttlepod provided 24 crew-days of breathable atmosphere. They were also provided with ration packs and a small protein resequencing unit.

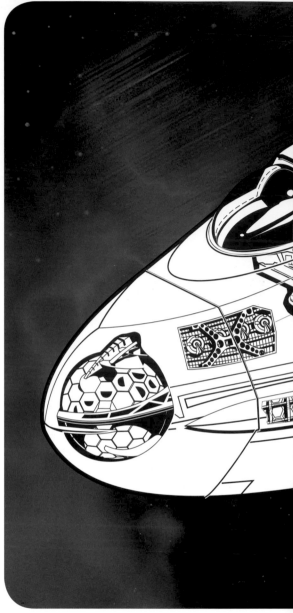

The kind of missions that required this sort of extended stay away from the ship included scientific analysis, for which the shuttlepods were equipped with quantum-level sensors.

The shuttlepods could also be used for military operations in or out of a planet's atmosphere and they were fitted with plasma cannons with a range of just under 10km.

In the event of a crash-landing, the shuttlepods carried emergency beacons with a range of ten million kilometers.

The NX-01's shuttlepods were based in a twin drop bay, which was depressurized when they were launched.

The shuttlepods were used to ferry crew members to and from the surface of planets and to gather readings in space.

Shuttlepod

Classification	OTV Type K42/Personnel
Active service	2151–2161
Length	6.9m
Crew complement	1 to 7
Propulsion	Nuclear fusion
Weaponry	Plasma cannons
Defenses	Polarized hull plating

Faster than light

WARP THEORY

The faster-than-light engine is arguably the most important invention in human history. The distances between stars are so great that, even traveling at the speed of light, it would take decades to reach the nearest inhabited star system. Zefram Cochrane's first warp flight in 2063 not only broke a massive technological barrier but also attracted the attention of the Vulcans, leading to the first official contact between humans and an alien race.

The physics behind warp drive are mind-bendingly difficult and can only really be explained with inadequate metaphors. For years, people believed faster-than-light travel to be impossible. This was famously set out in Einstein's theory of special relativity, which held that no matter with mass could be accelerated to the speed of light. This means that the fastest thing in the universe is something with no mass at all, such as a beam of light. And the speed of light in a vacuum has been measured at 299,792,458 meters a second, or 671 million miles an hour. At that speed it takes the light from the star nearest Earth, Proxima Centauri, approximately four years to reach Earth, making interstellar travel very, very difficult.

What Zefram Cochrane and his team realized was that Einstein's theory of general relativity said that space was curved rather than flat and that matter and energy could warp it. Since they couldn't make a ship go faster than light, what they had to do was warp space itself to make the distance between objects shorter. Imagine that space is a tablecloth. What a warp drive does is pull the bit of cloth in front of it up tighter together and then ride over the top of it. The cloth then gets pushed back to a normal, flat shape behind it. The ship itself is in a warp bubble, sometimes called a warp shell. Inside the bubble space isn't distorted and the ship is technically traveling at sublight speeds, but the bubble itself pushes through space faster than light can.

Doing this takes an enormous amount of energy. Cochrane's warp engine generates this by creating a matter/antimatter reaction that produces high-energy warp plasma. This reaction is controlled by dilithium crystals, which are a key to faster-than-light travel.

The super-heated plasma is then channeled to the ship's nacelles where it passes through the warp coils. These are made of a substance called verterium cortenide, and when the plasma passes through them they generate a warp field, literally changing the shape of the space around them.

The amount of distortion, and therefore the speed at which the ship is traveling, is measured on the warp scale. This has had to be redrawn in the early 24th century after it was demonstrated that warp-capable ships can travel far faster than anyone originally imagined.

However, the early development of the warp drive was surprisingly slow. Cochrane made his first warp flight in a converted Titan V missile (which he renamed the *Phoenix*) on April 5, 2063. This ship was capable of traveling at Warp 1, or the speed of light. Earth then made contact with the Vulcans, who by this point had had warp technology for centuries.

WARP PROPULSION

The warp nacelles on a starship generate a series of nested warp energy fields, which can be manipulated to control the speed of the ship. The fields create a warp bubble around the entire ship, but the amount of distortion varies according to the distance of the nacelle. The shape of the fields can be controlled from the front to the back of the nacelle by altering the frequency at which the plasma is sent through the verterium cortenide coils in each nacelle. The more the fields press against one another the more they distort the space around them and the faster the ship will move.

The exact shape of the warp fields is also influenced by the shape of the starship itself, which has been designed to correct the geometry and to ensure that the ship slips into warp easily.

Starships have two nacelles so that the shape of the field can be varied asymmetrically, thus allowing the ship to maneuver at warp speed. For example, the ship can turn to the starboard by making the field generated by the starboard nacelle proportionally weaker than the fields on the port side.

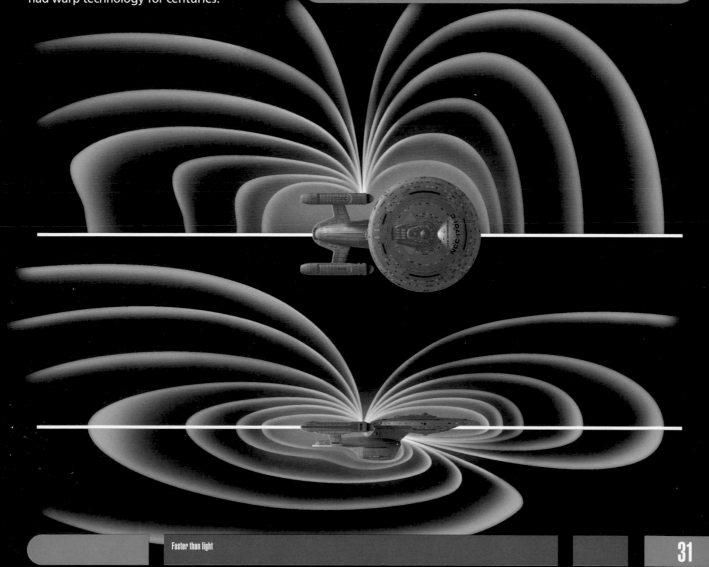

Although the Vulcans helped Earth develop many technologies and wiped out many problems such as hunger, they were very cautious about giving humans—who they regarded as very emotional and unpredictable—access to warp technology. Nevertheless, Earth continued to develop its own warp engines and by the 2140s a team led by Henry Archer had developed an engine that was theoretically capable of achieving Warp 5, or 125 times the speed of light.

Achieving Warp 5 was seen as a landmark, because it was the speed at which interstellar travel could be undertaken in hours or days rather than months or years.

The first Warp-5 capable ship, the *Enterprise* NX-01, was launched in April 2151 under the command of Henry's son, Jonathan Archer. The ship finally passed Warp 5 on February 9, 2152.

Archer's engine solved most of the serious problems confronting warp engineers and rapid progress was now made. Starfleet had

achieved Warp 7 by 2161, and by the time the *U.S.S. Enterprise* NCC-1701 launched in the 2240s ships had a cruising speed of Warp 6 and could achieve speeds as high as Warp 9. In fact, after being modified by the Kelvans in 2268 the *Enterprise* NCC-1701 actually achieved Warp 11 and would subsequently achieve Warp 14.1.

This led Starfleet engineers to redraw the warp scale. On their new scale Warp 10 represented an infinite velocity. If a ship ever traveled this fast it would literally occupy every place in the universe at once. (This was theoretically impossible but, as always, reality proved otherwise and a shuttle from the *Intrepid*-class *U.S.S. Voyager* NCC-74656 apparently achieved Warp 10 in 2372.) As the redrawn scale approaches Warp 10 the speeds get exponentially faster. In the 2370s starships could routinely achieve speeds above Warp 9.2, and even above Warp 9.9, but above this velocities became vastly greater for each point on the scale.

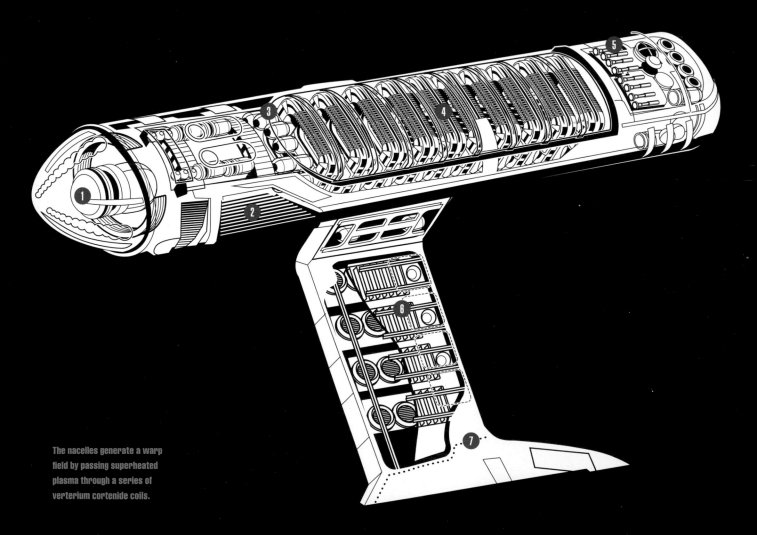

The nacelles generate a warp field by passing superheated plasma through a series of verterium cortenide coils.

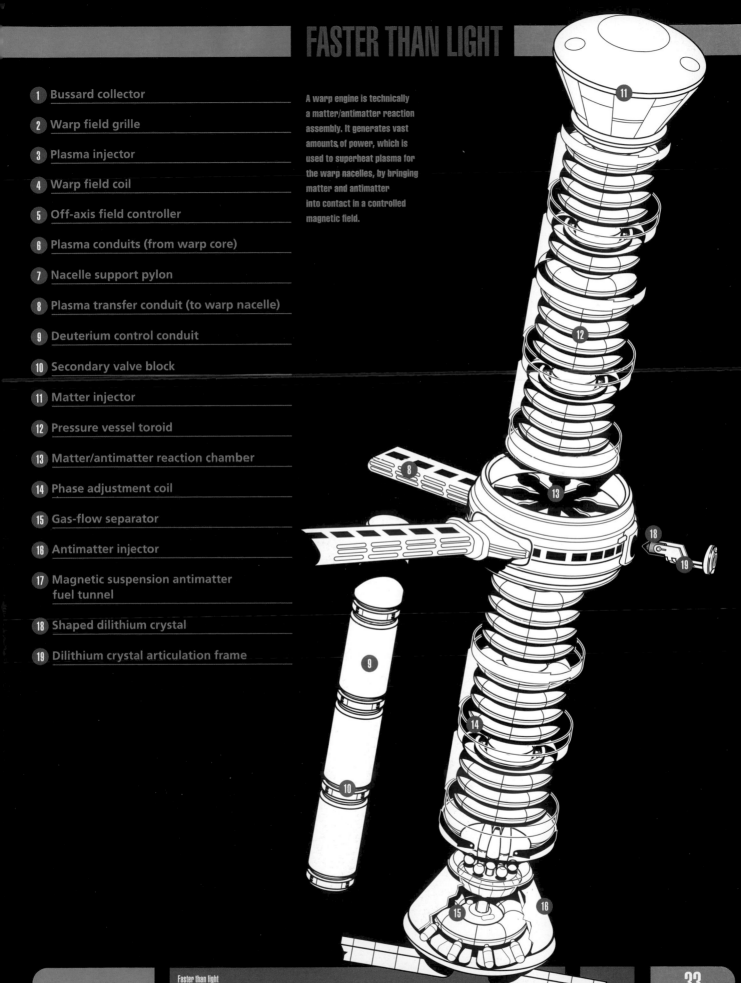

1. Bussard collector

2. Warp field grille

3. Plasma injector

4. Warp field coil

5. Off-axis field controller

6. Plasma conduits (from warp core)

7. Nacelle support pylon

8. Plasma transfer conduit (to warp nacelle)

9. Deuterium control conduit

10. Secondary valve block

11. Matter injector

12. Pressure vessel toroid

13. Matter/antimatter reaction chamber

14. Phase adjustment coil

15. Gas-flow separator

16. Antimatter injector

17. Magnetic suspension antimatter fuel tunnel

18. Shaped dilithium crystal

19. Dilithium crystal articulation frame

A warp engine is technically a matter/antimatter reaction assembly. It generates vast amounts of power, which is used to superheat plasma for the warp nacelles, by bringing matter and antimatter into contact in a controlled magnetic field.

Faster than light

ALTERNATIVE TECHNOLOGIES

Although warp drives based on Zefram Cochrane's original engine are still the most common way of traveling faster than light, they're certainly not the only way of reaching that kind of speed.

WORMHOLES

The fastest method of travel known to Starfleet is a wormhole. These are basically corridors through space that enable vast distances to be crossed almost instantaneously. The theory behind them is similar to that used in warp drive, only the distortions involved are much greater. Imagine that normal space is like a shoelace. If you travel along the lace it could take a huge amount of time to cover the whole distance; but a wormhole takes the two ends and puts them next to one another and allows you to jump across.

The Bajoran wormhole is the only stable wormhole that Starfleet has ever encountered. It links two parts of the Galaxy that are 70,000 light years apart.

Wormholes occur naturally in space although no one has ever discovered a stable naturally occurring wormhole. The entrances and exits flip wildly around the universe, moving on after normally as little as a few minutes. It was hoped that the Barzan wormhole (which was discovered in 2366) was stable and had terminuses in both the Gamma Quadrant and in the Alpha Quadrant near the planet Barzan, but it proved to be unstable and therefore worthless.

The only known stable wormhole in the Galaxy is the artificially created Bajoran wormhole, which links the planet Bajor with the Gamma Quadrant, approximately 70,000 light years away. The wormhole opens only when a vessel approaches it and although various ships had been lost near it, it was only 'discovered' in 2369 when Commander Benjamin Sisko and Lt Jadzia Dax stumbled across it. It was created by an ancient race known as the Prophets, who exist outside normal space-time.

ICONIAN GATEWAYS

An ancient race known as the Iconians developed a technology that allowed them to open what can only be described as doorways, which led to any point in the Galaxy. The science behind the gateways is far beyond the Federation and it is unclear how they worked. What is known is that the Iconians were destroyed some 200,000 years ago and their technology was lost. Two Iconian outposts have been discovered in modern times, one on the planet Iconia, the other on the planet Vandros IV in the Gamma Quadrant. Both outposts still had functioning gateway technology. A device in the outpost generated the doorway, which it seems could be opened to any point in the Galaxy. All a person had to do was step through. In both cases the Iconian outposts were destroyed to prevent them falling into the hands of the Federation's enemies.

THE SOLITON WAVE

In 2368 Dr Ja'Dar of Bilana III conducted experiments using soliton-wave technology that were designed to remove the need for warp engines. The theory was that a series of wave generators would create a self-reinforcing warp field that would envelop a ship and accelerate it to high warp speeds. The technology had some drawbacks—most significantly the wave could only travel between two fixed points. It was initiated by a soliton-wave generator and dispersed at the other end by a station that emitted a scattering field, to disperse the wave and ensure that anything traveling in it dropped out of warp. The advantage was that any ships traveling in the wave didn't require a warp engine. The system was also incredibly effective—the transfer of speed was up to 450% more efficient than a conventional warp engine. The experiments were only partially successful and Dr Ja'Dar encountered considerable difficulties maintaining a stable wave. It also proved very difficult to dissipate the wave, which continued to build in power. The wave would have destroyed everything in its path if the *Enterprise*-D hadn't managed to break it up by firing photon torpedoes into its path, causing a disruption in subspace.

TRANSWARP CORRIDORS

The Borg use a system of transwarp corridors that are technically superior to standard warp drives. By using them, the Borg can achieve speeds 20 times faster than a Starfleet vessel. Transwarp conduits are very closely related to wormholes and are effectively a fast-moving area of space—once a ship enters one it is accelerated to extremely high speeds. However, they have a short lifespan unless they are maintained. Borg ships are equipped with transwarp drives that can create such conduits. The conduits themselves are connected to form a network that allows the Borg to cross vast distances at incredible speeds. The network is maintained by six transwarp hubs. The destruction of one of these in 2379 had a severe impact on the Borg, crippling the network. Once a transwarp corridor has been created it has to be accessed. This is done by emitting a high-energy tachyon pulse, which opens the conduit.

QUANTUM SLIPSTREAM

Quantum Slipstream technology is closely related to transwarp but is even more advanced. It works by altering the quantum state of space to change the curvature of space-time. Unlike a warp drive it works by creating a narrowly focused warp distortion in front of the ship.

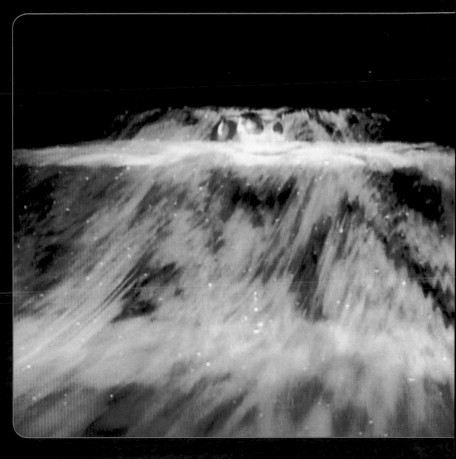

Ships could ride a soliton wave, like a surfboard.

Transwarp corridors are like a warp speed tunnel in space.

U.S.S. Enterprise
NCC-1701

The *U.S.S. Enterprise* NCC-1701 was the first Federation vessel to carry the famous name. She served with great distinction in the second half of the 23rd century, and was on active duty for longer than any of the other *Enterprises*. During her 40 years of service she was commanded by some of Starfleet's most famous captains including Christopher Pike and James T. Kirk. Her crew made major contributions to the Federation's understanding of the universe. The ship's mission to seek out new life and new civilizations took her beyond the edges of explored space and she visited both the center and the outer edges of the Galaxy. *Enterprise* was also engaged in some of the most important military operations of the era, which brought her into conflict with the Klingons and the Romulans.

Classification	Constitution class
Constructed	San Francisco Yards, Earth
Launch date	2245
Destroyed	2285 [autodestruct in orbit around the Genesis planet]
Length	289m; 305m [2271 refit]
Number of decks	23
Crew complement	203 [date: 2254]; 430 [date: 2264]
Weaponry	Phasers and photon torpedoes
Commanding Officers	Robert April, Christopher Pike, James Kirk, Willard Decker, Spock

CONSTITUTION CLASS

* The U.S.S. *Yorktown* was renamed the U.S.S. *Enterprise* NCC-1701-A following the destruction of *Enterprise* NCC-1701 in 2285.

In the second half of the 23rd century *Constitution*-class starships were in the front line of Starfleet's mission to explore the Galaxy. They were tasked with supporting newly established colonies and research missions, exploring new worlds, and seeking out new life. They also patrolled the Federation's borders, ready to respond to hostile alien powers. All *Constitution*-class ships served with distinction, but the *Enterprise* NCC-1701 became the most famous.

The ship was launched from Earth's San Francisco shipyards in 2245 under the command of Captain Robert April. She was subsequently commanded for over a decade by Captain Christopher Pike but earned her place in history after James T. Kirk assumed command in 2263. The following year Kirk embarked on a five-year mission that is among the most significant in the Federation's history.

Under Kirk, the *Enterprise* NCC-1701 had one of the most impressive crews Starfleet has ever assembled, and Kirk himself became one of the most decorated captains in Starfleet's history. His science officer and second in command, Spock, was the only Vulcan serving on a human Starfleet vessel at the time. The half-human, half-Vulcan officer would help to integrate Vulcan more closely with its human partners in Starfleet and would ultimately become one of the Federation's greatest ambassadors. The ship's chief medical officer, Dr Leonard H. McCoy, is famous as the author of *Comparative Alien Physiology*, a seminal textbook that is primarily based on material he gathered while serving on the NCC-1701. The ship's Chief Engineer, Montgomery Scott, is a legend in engineering circles and he virtually rewrote the book on starship operations.

During Kirk's first five-year mission the NCC-1701 and her crew were involved in many historically important events. Among their numerous scientific discoveries the crew probed beyond the edges of known space, developed a reliable method of time travel, and proved the existence of parallel universes.

Shortly after Kirk assumed command the *Enterprise* NCC-1701 was the first starship to survive an encounter with the great energy barrier that surrounds the Galaxy. Crossing the barrier nearly destroyed the ship, and its effect on the crew showed just how dangerous an encounter with the barrier can be, since it enhanced latent psychic powers in two crewmembers, Lt Gary Mitchell and Dr Elizabeth Dehner, giving them almost godlike powers and driving them mad.

Even more significantly, the *Enterprise* NCC-1701 was the first Federation vessel to document and prove the practicality of time travel. In 2267, following a serious infection that incapacitated the crew, the ship underwent an attempted cold restart of the warp engines using a new intermix formula that was developed by Chief Engineer Montgomery Scott and Science Officer Spock. The resulting warp

distortions and speed sent the *Enterprise* 71 hours back in time, proving beyond doubt that time travel was possible.

A few months later the NCC-1701 was involved in an accident with a black hole that sent her much further back in time, the ship being sent to the year 1969. The crew was able to return to the 23rd century by taking advantage of the massive gravitational forces generated by warp engines and a slingshot maneuver round the sun. This established a workable method of time travel that was used by Starfleet for many years.

In 2267 the *Enterprise* NCC-1701 proved the existence of parallel universes when an ion storm caused a transporter accident that sent the landing party to a 'Mirror Universe'. This universe was very similar to our own and yet different. Kirk's team managed to return by replicating the conditions that caused the accident.

The NCC-1701 was also involved with some of the most important military and diplomatic incidents of the time. The late 23rd century was an extremely volatile period and the Federation had hostile relations with the Romulans, the Klingons, and the Tholians. War was a constant danger and the *Enterprise* NCC-1701 was instrumental in avoiding at least two major conflicts. On Stardate 1709.2 she responded to a distress call from Federation outposts along the Romulan Neutral Zone. By the time the ship arrived, the outposts had been destroyed by a Romulan vessel using a cloaking device that made it almost impossible to detect. Kirk reasoned that the Romulans were testing the device and that if it returned unharmed to Romulan space the Federation's old enemy would almost certainly launch a full-scale war. The *Enterprise* therefore pursued the cloaked vessel and thanks to Kirk's tactical brilliance managed to destroy it.

Their cloaking technology gave the Romulans a major tactical advantage over the Federation and two years later, in 2268, *Enterprise* NCC-1701 was involved in a covert operation to steal a cloaking device that involved Kirk pretending to suffer a nervous breakdown and allowing his ship to cross into the Neutral Zone.

The danger of war with the Klingon Empire was even more acute. In 2267 peace negotiations between the Federation and the Klingons broke down, and the *Enterprise* NCC-1701 was sent to the strategically important planet Organia to offer its inhabitants the Federation's protection. The Klingons also arrived on the planet and the two powers were on the brink of all-out war when the Organians, an apparently technologically retarded civilization, revealed themselves to be extremely evolved and powerful beings who imposed a peace treaty that held for decades.

The *Enterprise* NCC-1701 completed her historic five-year mission in 2270, returned to the San Francisco spaceyards, and began a major refit that would see all of her systems significantly upgraded. Kirk accepted a promotion to admiral and took up a position as head of Starfleet operations. Spock and McCoy both left Starfleet, Spock to pursue the study of perfect logic on Vulcan and McCoy to enter private practice on Earth. The 18-month refit was carried out under the command of Willard Decker.

In 2267 the crew discovered the Guardian of Forever, a portal that appeared to allow people to travel to any point in time and space. However, using it proved very dangerous.

The *Enterprise* NCC-1701 was often involved in conflict with the Klingons.

PORT ELEVATION

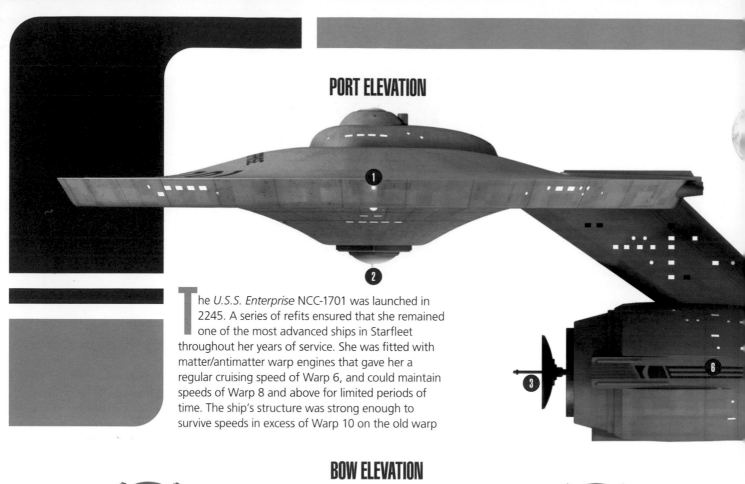

The *U.S.S. Enterprise* NCC-1701 was launched in 2245. A series of refits ensured that she remained one of the most advanced ships in Starfleet throughout her years of service. She was fitted with matter/antimatter warp engines that gave her a regular cruising speed of Warp 6, and could maintain speeds of Warp 8 and above for limited periods of time. The ship's structure was strong enough to survive speeds in excess of Warp 10 on the old warp

BOW ELEVATION

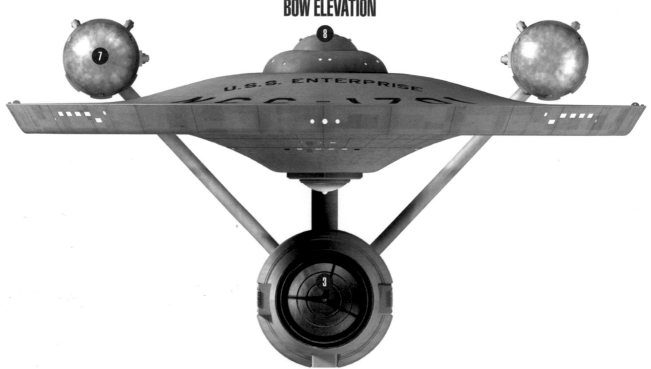

1	Saucer section	**4**	Warp nacelle
2	Sensor dome	**5**	Nacelle pylon
3	Long-range sensor/navigational deflector	**6**	Secondary hull

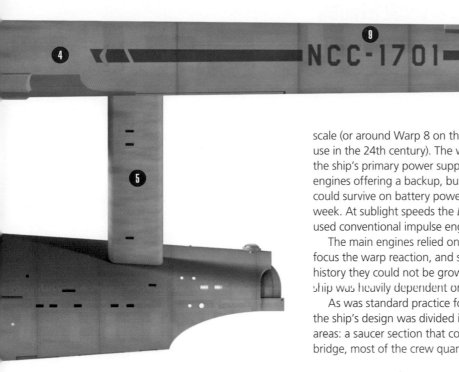

scale (or around Warp 8 on the redrawn scale in use in the 24th century). The warp engines were the ship's primary power supply, with the impulse engines offering a backup, but if they failed the ship could survive on battery power for approximately a week. At sublight speeds the *Enterprise* NCC-1701 used conventional impulse engines.

The main engines relied on dilithium crystals to focus the warp reaction, and since at this point in history they could not be grown or recrystalized the ship was heavily dependent on them.

As was standard practice for Starfleet vessels, the ship's design was divided into three distinct areas: a saucer section that contained the main bridge, most of the crew quarters, and the impulse

STERN ELEVATION

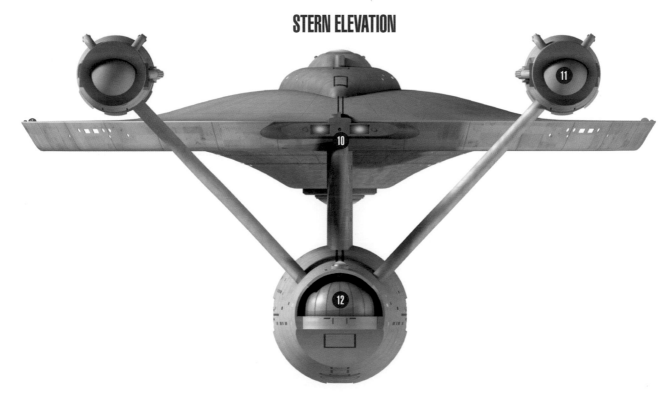

7 Bussard ramscoop

8 Bridge

9 Ship's registry

10 Impulse engines

11 Subspace field radiator

12 Shuttlecraft hangar deck

DORSAL ELEVATION

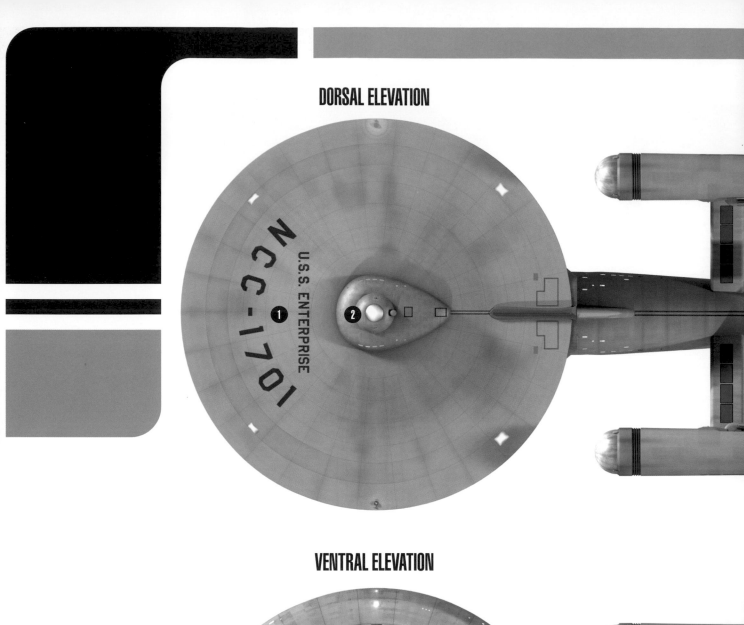

U.S.S. ENTERPRISE

NCC-1701

VENTRAL ELEVATION

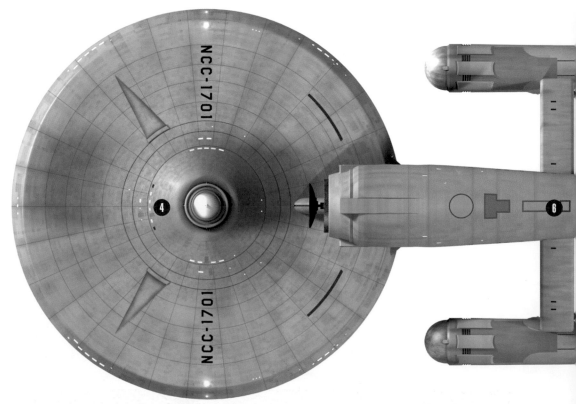

NCC-1701

NCC-1701

1 Ship's registry

2 Bridge

3 Warp intercooler intake

4 Phaser banks

5 Control reactors

6 Engineering hull

engines; an engineering hull that contained the warp drive systems, the shuttlecraft hangar deck, the main navigational deflector, and a wide variety of facilities; and twin nacelles that generated the warp field. In an emergency, these components could be separated from each other.

The saucer section was 11 decks deep, with the bridge on the top and the major sensor pod and phaser array on the underside. A refit in 2265 saw several significant modifications including a major upgrade of the bridge module, the removal of antennae from the front of the Bussard ramscoops, and a redesign of the rear of the nacelles.

The *Enterprise* NCC-1701 was a major spacegoing research facility and was equipped with 14 science labs, including stellar cartography and life sciences. Although Starfleet's primary mission has always been peaceful exploration, voyages into deep space are extremely dangerous, so the NCC-1701 was fitted with state-of-the-art weaponry. Photon torpedoes with a range of 750,000km could be fired from the underside of the saucer, which also housed massive phaser banks.

The ship was defended by deflector shields rather than the polarized hull plating of the previous century. These energy fields surrounded the ship and could protect it from repeated phaser blasts. If necessary, the strength of the field could be redistributed in order to provide additional protection to particular areas of the ship.

The shuttlecraft hangar deck was located at the rear of the engineering hull behind clamshell doors. *Enterprise* NCC-1701 carried a small complement of shuttles, each of which could carry up to seven people on short journeys. Under normal circumstances the shuttles were controlled by their own pilots, but the NCC-1701 was fitted with a powerful tractor beam that could be used to guide them, or other vessels, into the shuttlebay, to hold them in a stationary position relative to the ship, or to tow them.

Enterprise NCC-1701's shuttles were typically used for ferrying diplomatic passengers or for short-range scientific surveys. They were not needed for routine ship-to-planet missions since the ship was fitted with transporters that could beam personnel and equipment to and from a planet's surface.

The ship's computers used Richard Daystrom's revolutionary duotronic circuitry and the ship's hull was fitted with powerful sensor arrays that provided data to the science labs.

Since *Enterprise* NCC-1701's mission involved extensive periods away from Federation space, she was almost completely self-sufficient. The size of the crew varied significantly over the years. Under Pike the ship had a complement of just over 200, but by Kirk's time this had increased to 430.

MAIN BRIDGE LAYOUT

The main bridge of the *Enterprise* NCC-1701 was located at the very top of the saucer section on Deck 1. It took up the entire deck and was served by a single turbolift that delivered personnel to the rear of the room; a second turbolift was added in the 2270 refit. All the ship's major systems were controlled from the bridge, though in emergencies they could be accessed from an auxiliary control room located in the engineering hull.

The circular room was dominated by a main viewscreen. This was simply a computer monitor, the size of which varied considerably over the years: under Pike and at the beginning of Kirk's mission it was relatively small, but during a refit in 2266 it was replaced with a much larger version.

The majority of the consoles were positioned around the edge of the room on a raised platform, with the operators facing the walls. The captain's chair was in the center of the room, immediately behind the combined helm and navigation console.

Working clockwise from the main viewscreen the consoles were: defense subsystems, weapons subsystems, navigation subsystems, science, and communications. The science station, which incorporated the primary terminal for the ship's library computer, was normally operated by

The main bridge was the nerve centre of the *Enterprise*, and was normally manned by the captain and his senior staff.

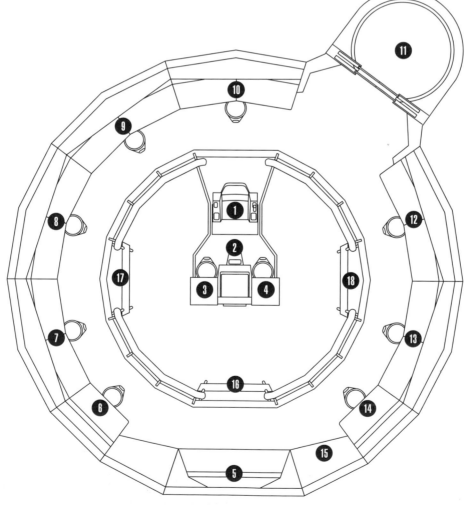

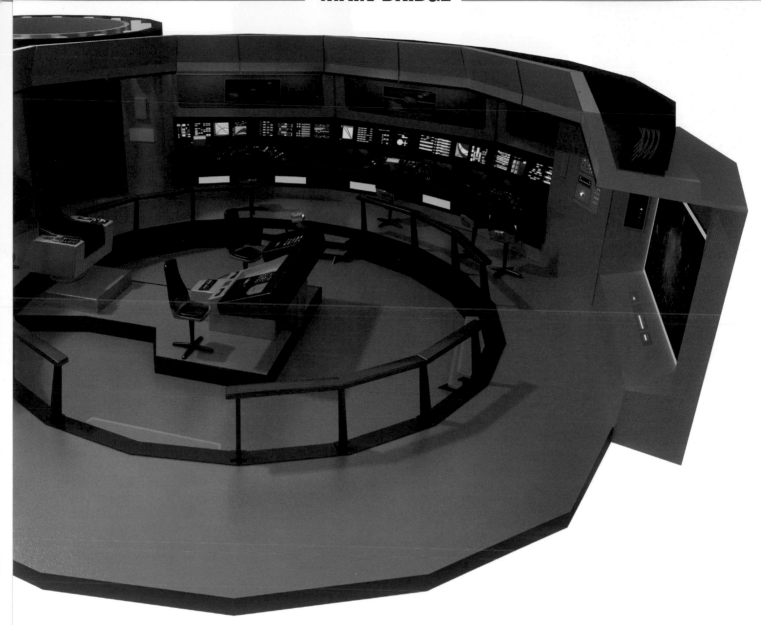

science officer Spock. The communications station was normally the responsibility of Lt Uhura. The stations on the other side of the turbolift, relating to engineering and environmental systems, weren't always manned during routine operations, their functions being primarily controlled from main engineering.

In an emergency, all bridge functions could be re-routed to main engineering or the auxiliary manual monitor control station.

THE CAPTAIN'S CHAIR

As is standard practice on Starfleet vessels, the captain's chair was located in the middle of the bridge, facing the main viewscreen. On the *Enterprise* NCC-1701 it could swivel, allowing the captain to face each station in order to hear his officers' reports. A number of important controls were built into

1. Captain's chair
2. Chronometer
3. Navigation systems
4. Helm
5. Main viewscreen
6. Defense subsystems
7. Weapons subsystems
8. Navigation subsystems
9. Science station
10. Communications station
11. Turbolift
12. Engineering station
13. Environmental subsystems
14. Engineering subsystems
15. Bridge support equipment
16. Steps to lower bridge level
17. Steps to lower bridge level
18. Steps to lower bridge level

HELM AND NAVIGATION

The twin console at the front of the bridge, known as the conn, was operated by a helmsman and navigator. The helmsman sat on the left and was responsible for piloting the ship, and was in direct communication with main engineering. To his or her right the navigator was responsible for monitoring the course and plotting any corrections. The two seats were divided by the ship's astrogator, which was used to input the ship's heading. Immediately below the astrogator, the ship's chronometer displayed the time—this was more important than one might imagine, since warp travel inevitably involves distorting the fabric of space-time and can cause significant problems with timekeeping.

The two officers shared tactical duties and had control of the ship's weapons and defensive systems. A targeting scanner was mounted in the helmsman's console. When needed this would be deployed and extended closer to eye level; the display provided analysis of vessels or phenomena outside the ship, including their composition, range, and status. It could also be used to target the ship's phasers and photon torpedoes. An indicator on the front of the console illuminated to show the ship's status. Under normal operating conditions it was not illuminated, but when the ship was at red alert it flashed red.

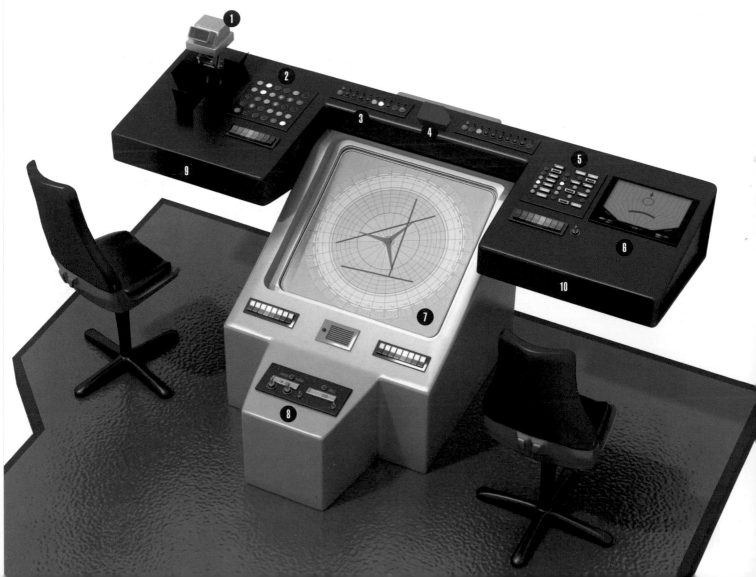

1. Targeting scanner
2. Helm control panel
3. Sensor activity monitors
4. Alert status indicator
5. Course programming panel
6. Deviation plotter
7. Astrogator
8. Ship's chronometer
9. Helmsman's station
10. Navigator's station

The original helm and navigation console (bottom) was later upgraded to include, among other improvements, the targeting scanner. This was normally concealed in the left side of the helmsman's station (middle), but could be deployed (top) for critical sensor analysis and tactical operations.

SPACE NAVIGATION THEORY

In starship flight operations, directions of external objects are generally specified in one of two ways: relative bearings and absolute headings. Relative bearings are specified in terms of angular difference—both azimuth and elevation—from the ship's nominal direction of flight. An object directly ahead of the ship would have an azimuth bearing of zero, while an object directly behind the ship would have an azimuth bearing of 180. Relative bearings are generally employed when describing the location of an object in relatively close proximity to the ship, such as an asteroid or another spacecraft. Absolute headings are specified in terms of angular difference from a vector drawn between the ship and the center of the galaxy. In this scheme, a vector directly toward the galaxy's center would be a heading of 000, mark 0, regardless of the orientation of the starship. Absolute headings are generally used in plotting interstellar navigation.

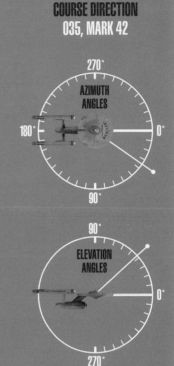

COURSE DIRECTION
035, MARK 42

AZIMUTH ANGLES

ELEVATION ANGLES

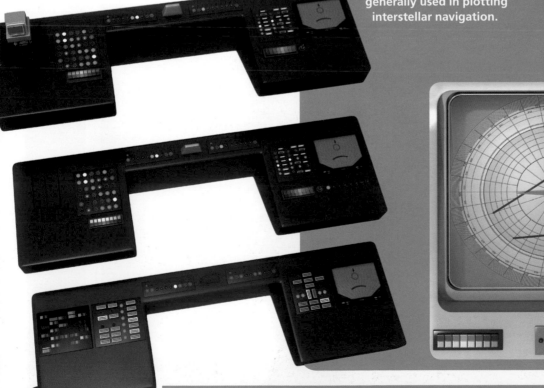

The *Enterprise* NCC-1701 was fitted with two main weapons—phasers and photon torpedoes; together these had the power to destroy the surface of an entire planet.

Under Captain Pike the ship was fitted with laser cannons, but by 2267 these had been replaced with more advanced phasers, which are a highly sophisticated form of directed energy beam. The word 'phaser' is actually an acronym for PHASed Energy Rectification. Banks of phasers were mounted on the underside of the ship's saucer section and had an effective range of 300,000km. They could be used in ship-to-ship combat or to target areas of a planet's surface from orbit. The intensity and function of the beam could be altered so that it could stun, heat, or disintegrate people or objects. An expert operator could use them to target an area as small as a building or to stun people in an area as wide as a city block. The ship's phasers were tied into the main engines and if necessary power could be diverted from them to the ship's shields. However, this would make the phasers less effective.

Because phasers are an energy beam they can only travel at the speed of light. In normal circumstances this makes them ineffective at warp speed.

The *Enterprise* NCC-1701 was also equipped with photon torpedo launchers. These torpedoes carried small amounts of matter and antimatter that caused a massive explosion when brought together. They had a small on-board engine that gave them an effective range of 750,000km, and, unlike phasers, they could be fired at warp speed. However, if a targeted ship responded quickly it could outrun a photon torpedo.

The *Enterprise* NCC-1701 was protected by deflector shields (also known as screens). The system created an energy field that could be projected around the ship, deflecting energy from enemy weapons and deflecting debris. The deflectors weakened every time they were hit and would fail under repeated bombardment. However, they were tied into as many power sources as possible and so energy could be redirected to them as necessary. The deflector shields were divided

The phaser has been the weapon of choice on Starfleet vessels since the 23rd century. They use phased energy beams to disrupt the structure of the target. The *Enterprise* NCC-1701's main phaser banks were located on the underside of the saucer by the sensor dome.

PHASERS

The phaser is a form of directed energy beam first developed in the mid-22nd century. It works by using something known as the Rapid Nadion Effect (RNE) and superconducting fuhigi-no-umi crystals; the version used in phasers is known as LiCu 518 crystals. Rapid Nadions are subatomic particles with a short life. When they are passed through the crystal they liberate and transfer strong nuclear forces creating the phaser beam.

The phaser is made up of several components. First of all electroplasma is channeled into a plasma distribution manifold (PDM). This splits the energy into separate pre-fire chambers made of LiCu 518 crystals. At this point the RNE causes a spectrum shift, which converts the plasma into a high-energy beam. This is passed to the beam emitter, which is a three-faced LiCu 518 crystal.

The power of the phaser beam is determined by the amount of plasma used, and the direction, width and intensity of the beam is determined by directing the energy into a different number of pre-fire chambers, each of which feeds a different part of the emitter.

Phaser power is extremely variable and can produce effects from light stun through to explosive, disruption, and even total disintegration.

into rear and forward groups and power could be switched between the two. The shields could also be extended around other, small vessels. In the 23rd century, when the shields were active the ship's transporters could not beam through them.

The deflectors could also be used as a form of cloaking device, making the ship invisible to primitive sensors. However, this approach was not effective against contemporary vessels such as those operated by the Klingons and the Romulans.

MARK VI-B PHOTON TORPEDO

1. Sustainer engine exhaust port

2. Service access port

3. Primary targeting scanner

4. Guidance and targeting processor

5. Wide field sensors

6. Intermix reaction chamber

7. Magnetic isolation pods

8. Sustainer propulsion module

9. Propulsion and thrust vector subsystems

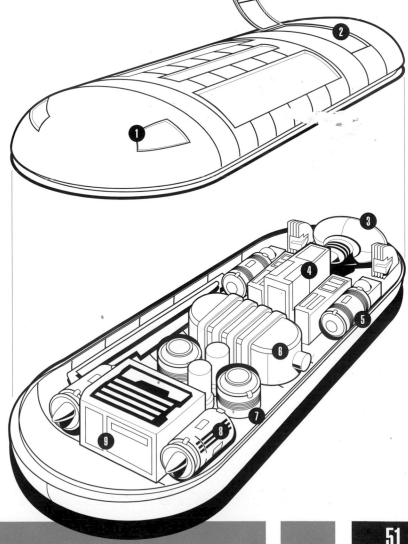

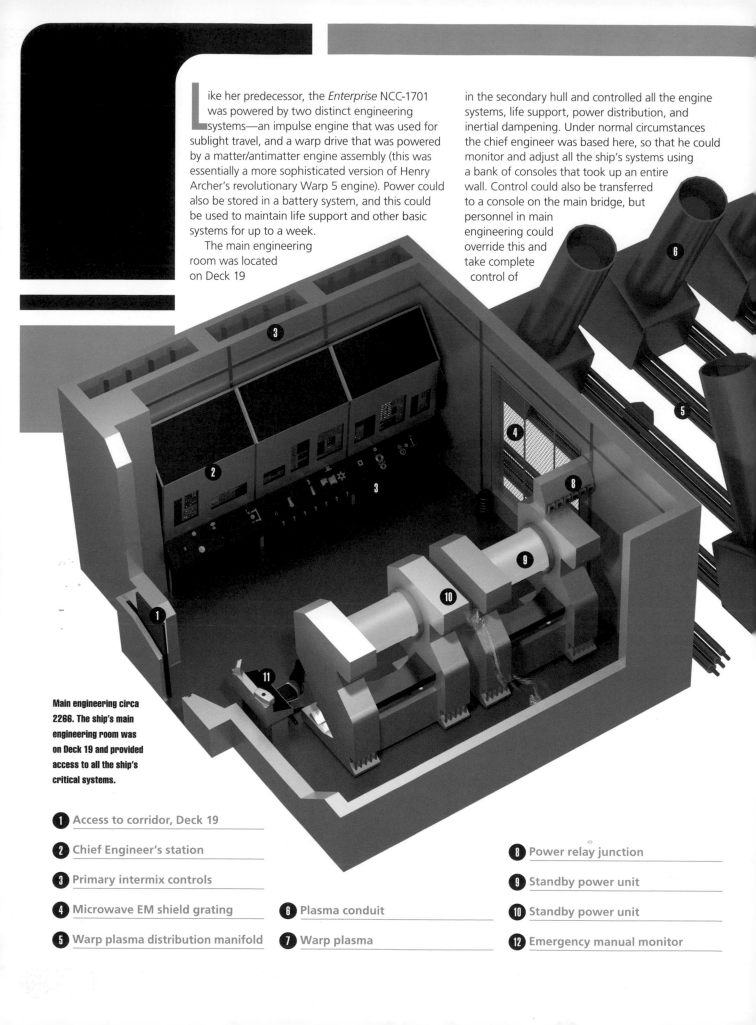

Like her predecessor, the *Enterprise* NCC-1701 was powered by two distinct engineering systems—an impulse engine that was used for sublight travel, and a warp drive that was powered by a matter/antimatter engine assembly (this was essentially a more sophisticated version of Henry Archer's revolutionary Warp 5 engine). Power could also be stored in a battery system, and this could be used to maintain life support and other basic systems for up to a week.

The main engineering room was located on Deck 19 in the secondary hull and controlled all the engine systems, life support, power distribution, and inertial dampening. Under normal circumstances the chief engineer was based here, so that he could monitor and adjust all the ship's systems using a bank of consoles that took up an entire wall. Control could also be transferred to a console on the main bridge, but personnel in main engineering could override this and take complete control of

Main engineering circa 2266. The ship's main engineering room was on Deck 19 and provided access to all the ship's critical systems.

1 Access to corridor, Deck 19

2 Chief Engineer's station

3 Primary intermix controls

4 Microwave EM shield grating

5 Warp plasma distribution manifold

6 Plasma conduit

7 Warp plasma

8 Power relay junction

9 Standby power unit

10 Standby power unit

12 Emergency manual monitor

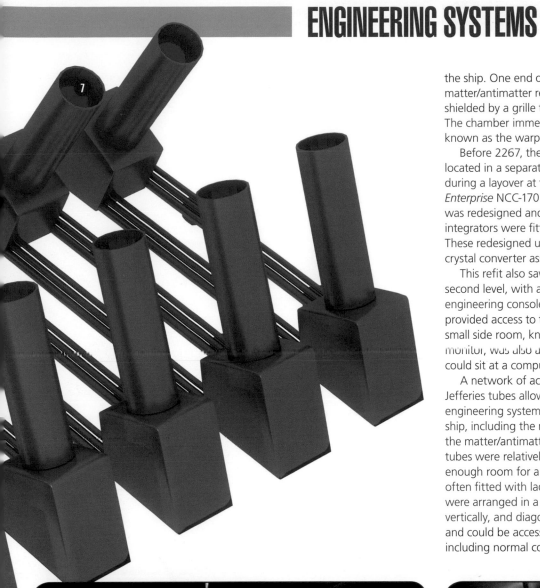

the ship. One end of the room was open to the matter/antimatter reaction assembly, which was shielded by a grille that filled most of one wall. The chamber immediately behind the grille was known as the warp power distribution manifold.

Before 2267, the dilithium crystals were located in a separate control room. However, during a layover at the end of 2267 the *Enterprise* NCC-1701's main engineering room was redesigned and the twin matter/antimatter integrators were fitted in the centre of the room. These redesigned units contained the dilithium crystal converter assembly.

This refit also saw main engineering gain a second level, with a new stairway between the engineering consoles leading to a gangway that provided access to the warp-drive systems. A small side room, known as the emergency manual monitor, was also added where the duty engineer could sit at a computer console.

A network of access shafts known as Jefferies tubes allowed personnel to access the engineering systems that ran throughout the ship, including the main junction circuitry and the matter/antimatter reaction chamber. Jefferies tubes were relatively cramped and provided only enough room for a single technician; they were often fitted with ladders or stairs. The tubes were arranged in a network that ran horizontally, vertically, and diagonally throughout the ship and could be accessed from a variety of points, including normal corridors.

Main engineering underwent a significant refit in 2267 when the dilithium-focused matter/antimatter integrators were moved to the center of the room and a second level was added.

A series of Jefferies tubes ran through the rest of the *Enterprise*, providing the engineering staff with access to a wide variety of systems. They were often used to reroute systems or patch circuitry when the ship was damaged in combat.

By the 2260s the transporter was the standard method for sending personnel from the ship to a planet's surface or other nearby vessels. The *Enterprise* NCC-1701's transporter room housed a large circular transport chamber. The floor of the chamber was a raised dais on which were located six personnel transport pads, arranged in a circular pattern. It was operated from a freestanding console by a transporter chief, or, in many cases, the chief engineer Mr Scott.

The transporters of this era had an effective range of 16,000km. In order to beam a crew member to or from a planet's surface, the transporter needed precise coordinates of the target subject or destination. These coordinates were normally provided by the target's communicator, though if necessary the transporters could be adjusted for a wide-field pickup that enabled them to beam up all the people in a given area.

When transporting to space stations, or other Starfleet facilities, it was standard practice to tie the transporter into another transporter pad at the other end, reducing the already small risk of accident. By the 2260s the *Enterprise* NCC-1701's transporters were sophisticated enough to beam people from the transporter room to another part of the ship.

The *Enterprise* NCC-1701's landing parties normally used the transporter room to beam down to the surface of a planet. Communicators were used to provide a signal lock so that people could be beamed back safely.

1 **Pattern buffer status indicators**

2 **Communications panel**

3 **Heisenberg compensator status indicators**

4 **Autosequence status display**

5 **Manual subsystem select**

6 **Energizer controls**

7 **Targeting scanner**

8 **Doppler compensation monitor**

1. Transporter control console

2. Transporter pad

3. Phase transition coil

4. Beam emitters

The *Enterprise* NCC-1701 had a single shuttlecraft hangar deck that was located at the rear of the engineering hull. Refits would add docking ports at other points around the ship, including the rear of the saucer section and the side of the engineering hull.

The distinctive clamshell doors of the original hangar deck opened directly on to space to allow shuttles to enter and exit the ship.

In order for a shuttle to be launched the hangar deck had to be completely depressurized. This meant it had to be evacuated and operations were coordinated between pressurized hangar deck control booths (on the upper level of the hangar deck) and the bridge. The shuttle's pilot was responsible for launching or landing the ship, though he was assisted by tractor beams, which prevented a collision. If necessary the entire landing process could be controlled by tractor beams.

One of the *Enterprise* NCC-1701's shuttles was positioned on a rotating landing pad. The chamber was depressurized, then the clamshell doors were opened, allowing the shuttle to leave the ship.

Constitution-class ships carried seven-person shuttles that were capable of interplanetary journeys. Unlike the *Enterprise* herself, they were designed to enter a planet's atmosphere and were capable of planetary landing and take-off.

Shuttles were used for scientific research missions, short-range transport missions, and ferrying diplomatic passengers to and from a planet's surface—for example, on the way to the Babel Conference in 2268 the *Enterprise* NCC-1701 received the Vulcan delegation on the hangar deck rather than in the transporter room.

The NCC-1701's shuttlecraft included the *Galileo* NCC-1701/7 – which was lost after it crash-landed on Taurus II, but was replaced with another shuttle that carried the same name – and the *Columbus* NCC-1701/2.

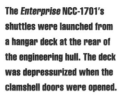

The *Enterprise* NCC-1701's shuttles were launched from a hangar deck at the rear of the engineering hull. The deck was depressurized when the clamshell doors were opened.

NCC-1701 / 6

Classification	Class-F
Active service	2267–2273
Length	7.3m
Crew complement	7
Propulsion	Ion engine
Weaponry	None
Defenses	Deflector shields

The *Enterprise* NCC-1701's main sickbay, located on Deck 7 in one of the most protected areas of the ship, acted as a surgery, a recovery ward, and a medical research facility. The ship was provided with four wards and a variable pressure chamber that could be used to expose crewmembers to different atmospheric pressures. This was sometimes necessary in order to prevent the bends, a condition that results from sudden decompression in an oxygen–nitrogen atmosphere. The chamber could also be used for hyperbaric and reduced-gravity therapy.

Main sickbay was divided into four distinct areas—a lab, the Chief Medical Officer's office, a diagnostic examination room, and a surgical and recovery ward. The lab and CMO's office were provided with full library access and contained a variety of analytical medical instruments. It was here that Dr McCoy did much of the research for his *Comparative Alien Physiology.* The next section of sickbay was given over to a variety of diagnostic instruments, including a full diagnostic biobed and a shorter examination bed equipped for cardiac stress testing. If a crewmember lay on the bed and used the pedals above it in the wall, the systems could record their heart rate and physical conditions during exercise. The diagnostic biobed had the standard life-signs monitor over the patient's head, which provided a constant stream of information on their heart rate, respiration, blood pressure, neural function, and other essential information. The panel could be recalibrated for patients of different species.

The diagnostic biobed could be rotated through 90° to

1. **Variable pressure chamber**

2. **Library computer access terminal**

3. **Electron microscope**

4. **Chief Medical Officer's office**

5. **Diagnostic biobed**

6. **Physical examination bed**

7. **Diagnostic examination room**

8. **Access to corridor, Deck 7**

9. **Alert status annunciator**

10. **Drug analyzer and synthesizer**

11. **Life-signs monitor**

12. **Surgery and recovery ward**

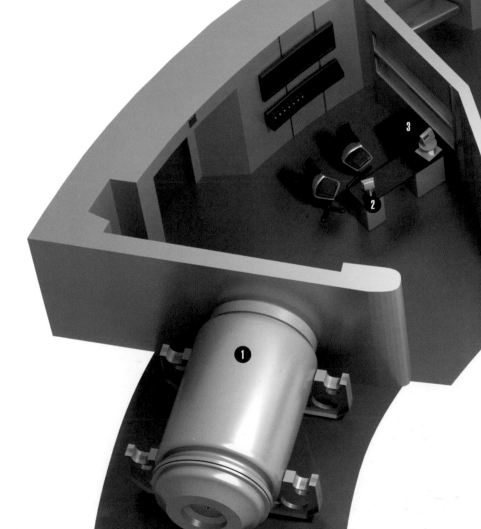

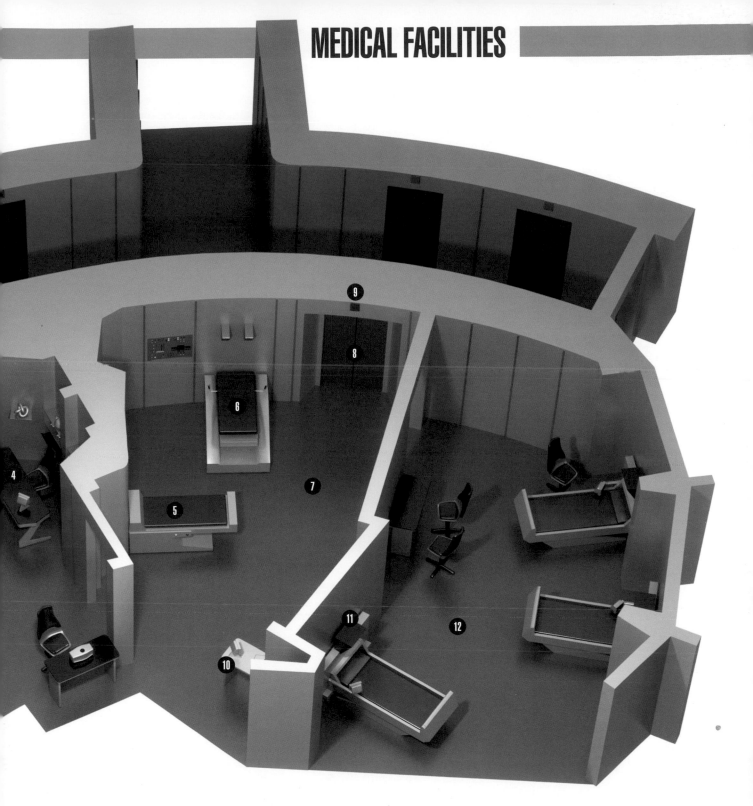

a standing position. It provided incredibly detailed information on a patient's condition and could also be used to perform autopsies.

A separate room served as both a surgery and a recovery ward. During major surgery—such as the operation on Ambassador Sarek's heart in 2267—a surgical frame was attached to the bed, over the patient, allowing the doctor to operate in a sterile environment, to provide precise information on the patient's condition, and to perform micro and

cryogenic surgery. The sterile field generated by the unit meant that surgeons could operate without gowns or masks.

The beds in the recovery ward all featured life-signs monitors that were mounted over the patient's head. Restraints were built into the sides in case patients became violent or suffered from dangerous fits. A swinging arm next to the bed was equipped with a library computer access terminal. This allowed recovering patients to read or study.

Dr McCoy's sickbay served as everything from a GP's office to a full-scale operating theatre. It was also a sophisticated medical research facility.

The majority of the ship's living quarters were located in the saucer section, with the senior officers' and VIP guest quarters on Decks 4, 5, and 6. Every crewmember on the *Enterprise* NCC-1701 was provided with their own quarters, which consisted of a single room separated into two distinct areas. The quarters could be configured according to the crewmember's personal preferences but always consisted of a sleeping area and a work area; a small bathroom with a shower was located off the sleeping area. The work area featured a desk with a computer access terminal. Each crewmember was provided with a safe in the partition behind the desk that could be accessed by touching a series of buttons over the top in the correct sequence. During the 2250s Captain Pike had larger quarters, though Captain Kirk favored standard crew quarters.

There were various communal areas throughout the ship including a rec room where crewmembers could play music or three-dimensional chess, a commissary (at this point in Starfleet's history ships still had fully crewed galleys), a gymnasium, and a chapel. All these facilities were significantly upgraded in the 2270 refit, at which point the crew quarters were rearranged so that junior crewmembers shared rooms.

1 Regular crew quarters

2 Recreational room

3 Chapel

4 Meeting room

5 Botanicals

6 Gymnasium

NCC-1701 [2270–2271 Refit] & NCC-1701-A

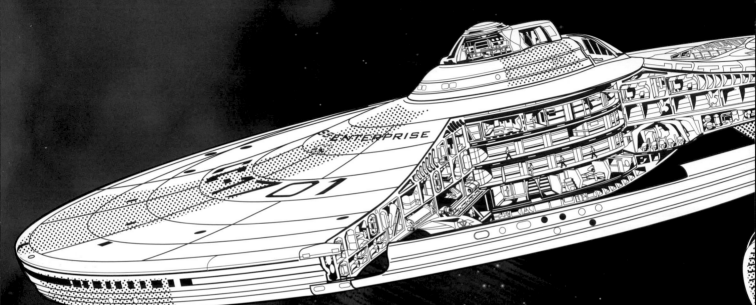

Classification	Constitution class (refit)
Launch date	2271 (1701); 2286 (1701-A)
Decommissioned	2285 (1701); 2293 (1701-A)
Length	305m
Number of decks	21
Crew complement	450
Weaponry	Phasers, photon torpedoes
Commanding Officers	Willard Decker, James T. Kirk, Spock

U.S.S. Enterprise
NCC-1701 [2270—2271 Refit]
NCC-1701-A

In 2269 the *U.S.S. Enterprise* NCC-1701 returned to Earth and underwent a major refit under the command of Captain Willard Decker. The now-Admiral Kirk resumed command in 2271 to deal with the V'Ger crisis, a mission during which Decker was lost in action. After this the *Enterprise* NCC-1701 (refit) became Kirk's flagship and was engaged in further deep-space missions before she was assigned to Starfleet Academy as a training vessel under the command of Captain Spock. Kirk once again took command in 2285 when Khan Noonien Singh stole the Genesis device—an experimental technology that had the capacity to destroy a planet. The refit *Enterprise* was seriously damaged when the Genesis device was detonated in the Mutara Nebula, and was subsequently destroyed in orbit around the Genesis planet. Another *Constitution*-class ship, the *U.S.S. Yorktown* NCC-1717, was recommissioned as the *Enterprise* NCC-1701-A in 2286 and eventually retired from service in 2293.

The refit *Enterprise* NCC-1701 was rushed back into service several months before the work was due to be completed in order to deal with the V'Ger crisis. V'Ger appeared to be a powerful energy cloud that was approaching Earth, destroying everything in its path, including the Federation's Epsilon IX monitoring station and three Klingon battlecruisers. The *Enterprise* was the only Federation ship capable of intercepting it. Given his years of experience and the seriousness of the threat, Admiral Kirk assumed command, making Decker his executive officer. The refit *Enterprise* was far from ready to enter full service but Kirk succeeded in intercepting the energy cloud and establishing that it was in fact a huge living machine that had once been the Earth probe *Voyager VI*.

V'Ger had attained sentience and had returned to Earth in order to find its creator. Decker and the *Enterprise*'s navigation officer Lieutenant Ilia merged with V'Ger, allowing it to appreciate the full range of human experience. The living machine somehow gained a new understanding of reality and literally vanished, presumably leaving our universe to explore new and different forms of existence.

Kirk now assumed permanent command of the refit *Enterprise* NCC-1701, making her his flagship, and embarked on another five-year mission. When this was completed in 2277 he accepted a posting at Starfleet Academy and the refit *Enterprise* became a training vessel under the command of the newly promoted Captain Spock. In 2285 Kirk took the ship, which was crewed by cadets from the academy, to investigate a loss of communication with the Regula I space laboratory, which was working on the Genesis project. This was an experimental terraforming device that was designed to transform uninhabitable planets into worlds capable of supporting life. Unfortunately it had the side effect of destroying all existing life, making the Genesis device a powerful weapon.

Kirk discovered that the device had fallen into the hands of the genetically engineered Khan Noonien Singh. After a battle in deep space the Genesis device was detonated in the Mutara Nebula, killing Khan and creating an entirely new planet. The refit *Enterprise* was severely damaged during the

After the end of Kirk's first five-year mission the *Enterprise* NCC-1701 underwent a major refit in orbit around Earth. She was barely spaceworthy when she was hurried into service to deal with a powerful entity known as V'Ger.

encounter and Captain Spock was killed. Starfleet Command decided to retire the ship and reassign the crew. However, Kirk was determined to retrieve Spock's body from the Genesis planet and return it to Vulcan. Since the Genesis planet had been placed off-limits, he and his senior staff stole the refit *Enterprise* and made their way to Genesis. When they arrived they discovered the Klingons were already there and Kirk was forced to destroy the *Enterprise* to prevent her falling into their hands.

Kirk and his crew eventually returned to Earth with a regenerated Spock, who had been miraculously restored to life by the Genesis process. In the course of their return they saved Earth from an incredibly powerful alien vessel and in recognition of this Kirk was given command of another *Constitution*-class ship. He was, however, demoted to captain for his part in stealing the refit *Enterprise* in direct contravention of his orders.

Kirk's new ship was the *U.S.S. Yorktown*, which had recently undergone a substantial refit. In honor of her predecessor she was recommissioned as the *U.S.S. Enterprise* NCC-1701-A. The use of a suffix in the ship's registry is a rare honor that is only granted to Starfleet's most distinguished vessels. It's a privilege that the *Enterprise* would retain, with the next four vessels of this name all using the NCC-1701 registry, even though they belonged to different classes of starship.

The new *Enterprise*-A completed her shakedown cruise in 2287 and began another deep-space mission under Kirk's command. She was finally retired in 2293, after being involved in the historic

negotiations with the Klingons at the Khitomer peace conference. Appropriately, the *Enterprise*-A was instrumental in destroying a cloaked Klingon vessel commanded by the Klingon General Chang, who was intent on disrupting the peace process. Kirk and Spock also exposed Starfleet officers who were equally determined to avoid peace with the Klingons.

The *Enterprise*-A served until 2293, after her crew helped open negotiations with the Klingons, which finally led to the Khitomer Accords.

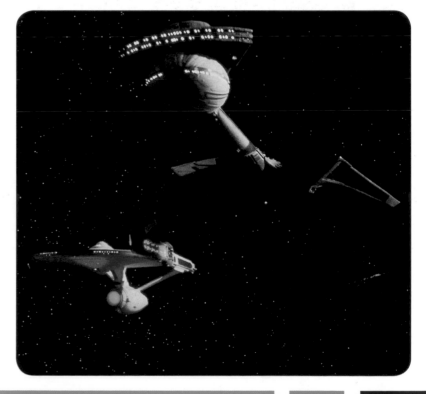

How transporters

MATTER STREAM THEORY

The transporter is one of the most significant inventions in Earth's history. It has become the standard means of transportation, vastly reducing the need for cars and roads, and has had a profound influence on city planning and architecture. The same technology was used in replicators, which are routinely used to manufacture everything from spares to houses, and revolutionized farming by providing meat without slaughtering animals. In terms of space exploration it made a journey from a starship to a planet's surface into a simple operation that took a matter of seconds.

The transporter was invented by Emory Erickson in the early years of the 22nd century but wasn't used routinely until the 2250s. When humans first started to explore the Galaxy transporters were only suitable for sending non-living materials such as food and supplies. *NX*-class ships such as the *Enterprise* NX-01 were among the first starships to be fitted with transporters authorized to transport living beings. Nevertheless, the crew were extremely wary about using them and not without good cause. Using a transporter involves literally disassembling a person atom by atom, converting them into a beam of energy, sending this through space, and then reassembling them at the other end.

Transporters are extremely reliable and are used to make millions of journeys every day, but there are clearly a lot of things that can go wrong with the process—the beam can get degraded, and if it is reassembled in even slightly the wrong place it could transport someone into the ground, causing an excruciating death. This is without even addressing the metaphysical questions of what happens to a person's soul during this process. Even 100 years after transporters started to be used, many people, notably Dr McCoy, were extremely uncomfortable about using them.

TRANSPORTER OPERATION

In the simplest terms a transporter analyzes the person or object to be transported at a subatomic level, recording the exact position and energy state of

work

1 **Operator's console**

2 **Primary energizing coils**

3 **Molecular imaging scanner**

4 **Transporter pad**

5 **Phase transition coils**

6 **Pattern buffer**

7 **Biofilter**

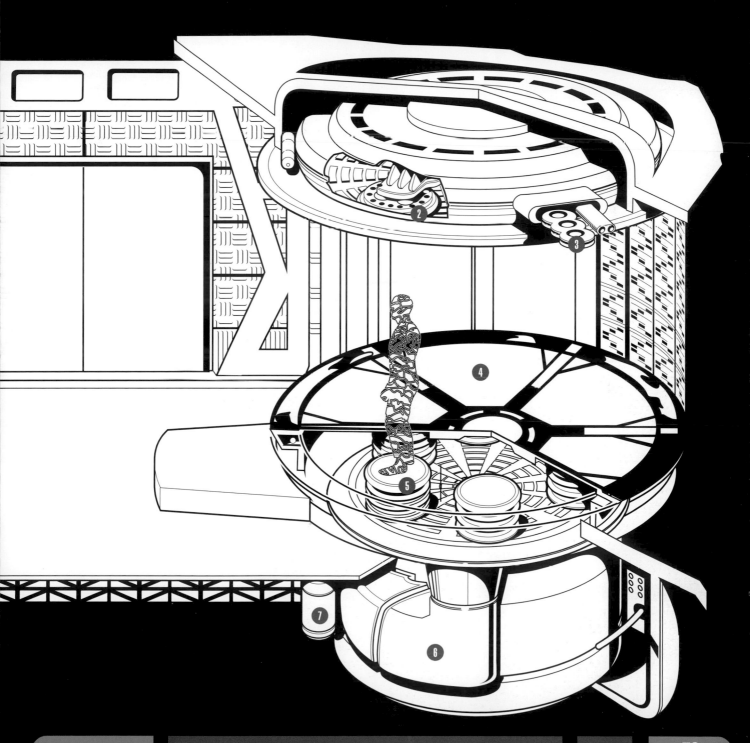

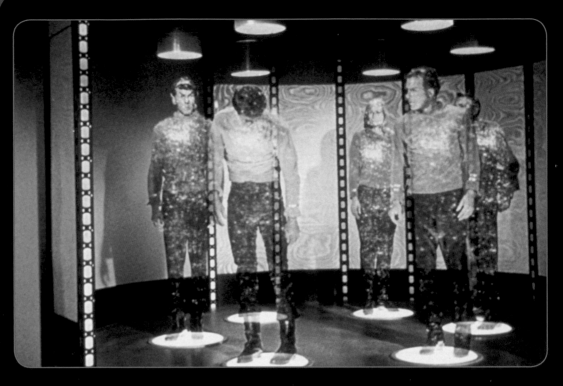

Anyone or anything standing on the *Enterprise*'s transporter pad could be broken down into a stream of matter and energy that could then be beamed to a different location.

every subatomic particle at a quantum level. The person is then converted into a beam of energy, or 'energized'. Technically this is a subatomically debonded matter stream.

This matter stream is then held in a 'pattern buffer'. While the beam is here the computers are able to compensate for any differences in relative velocity (in essence the different speed of movement) between the transport chamber and the destination.

The matter stream is then fed to one of the

transporter emitters on the outside of the ship and beamed to the destination, where it is reassembled.

In theory if anything goes wrong, the buffer can be used to reassemble the person in the transporter. In practice this is a question of how far into the process the problems occur.

It's possible to hold a person in the pattern buffer for a limited period of time, though the quality of the image normally starts to degrade after a few minutes. Chief Engineer

TRANSPORTER OPERATION TIMELINE

00.000	00.001	00.007	00.046	00.057	00.063	00.072	00.085
Autosequence initiation	Auto-diagnostic sequence begins	Transport system query for verification of signal routing, pattern buffer availability, and backup pattern buffer assignment	Diagnostic verification of controller logic states	Diagnostic verification of targeting scanners and Doppler compensation	Diagnostic verification of pattern buffer	Diagnostic verification of backup pattern buffer	Pattern buffer, and backup pattern buffer initialization

00.151	00.236	00.259	00.327	00.332	00.337	00.338	00.341
Energise to 1.7 MeV. Initial frequency set at 10.2 Ghz	Superconducting tokamak to operating capacity	Energise ACB elements to 1.7 MeV	Syncronize pattern buffer with phase transition coils	Reset molecular imaging scanners to null	Quark resolution enhancement enabled on molecular imaging scanners	ACB elements to initial operating level of 12.5 MeV	Operating panel indicates commencement of energizing sequence to be controlled manually at operator's discretion

Montgomery Scott managed to find a solution to this and when his ship, the *Jenolen*, was nearly destroyed in 2294 he managed to store himself in his transporter's pattern buffer for 75 years until he was discovered by the crew of the *Enterprise*-D. He stopped his pattern from degrading by locking the unit into diagnostic mode, so that the matter array was cycled around the pattern buffer, and cross-connecting the phase inducers to provide a regenerative power source. Because he was stored as pure energy he didn't age or need food.

TECHNOLOGICAL ADVANCES

Over the years transporters have become much more sophisticated. The very first transporter took 90 seconds to cycle through and people being transported could feel themselves being literally disassembled.

Early models were not sophisticated enough to transport between vessels that were moving at high velocities. This limitation was overcome by Commander Scott when he realized that the answer was to treat space itself as a moving object. Even so it was extremely difficult to transport anything through a ship's deflector shields. This isn't surprising, since the whole purpose of the shields is to deflect energy. This has meant that transporters have very limited uses in combat. Not only can a ship not transport people on to an enemy vessel, but if the captain wants to beam their own personnel back on to the ship they will have to drop the shields, leaving the ship vulnerable.

By the late 24th century this limitation had been partially overcome by closely aligning the frequencies of the shields and the transporter beam. If someone knew the frequency of the shields they could adjust the transporter to beam through them.

For a long time it was believed that it was impossible to transport between two ships traveling at warp speed. This isn't just about dealing with two objects traveling at incredible velocities—the transporter also has to compensate for the spatial distortions caused by the warp fields. A method was devised by the late 24th century, but it is not advisable and is only to be attempted in emergencies.

Over time Starfleet's engineers also learned how to modify the contents of a transporter beam before reassembling it. By the 23rd century, this was done routinely to filter out any bacterial or viral infections that a crewmember might have picked up on a planet's surface, thus removing the need for the decontamination chambers that were essential on early starships. This technology was known as a biofilter. The system isn't flawless, however, since the filters can't always identify contaminants.

The same basic technology can also be used to remove weapons from a person or even to remove a weapons discharge from the transporter beam so that it doesn't rematerialize at the destination.

Transporter technology has also been put to good use in other areas. The same basic process can be used to create perfect copies of most non-living objects. This revolutionized the

00.097	00.102	00.118	00.121	00.138	00.140	00.142	00.145
Diagnostic verification of phase transition coils	Reference signal activated	Verification of emitter array assignment	Diagnostic verification of emitter arry and waveguide conduits	Diagnostic verification of molecular imaging scanners and Heisenberg compensators	Target scan verification of beamdown coordinates	Operating panel indicates system readiness	Begin emission of annular confinement beam in chamber

00.359	00.363	00.417	00.432	00.464	00.523	00.596	00.601
Molecular imaging scanners commence scan sequence. Reference beam frequency lock	ACB modulation lock	Phase transition coils begin ramp-up to 162.9 Ghz. Energize to 32 MeV	Molecular imaging scanners begin transmission of analog image data to pattern buffer	Verification of image data integrity	Frequency syncronization of pattern buffer with phase transition coils	Phase transition coil frequency locked to 162.9 Ghz. Commence dematerialization cycle	Transport ID trace stored to provide record of transporter activity

manufacture of countless materials and equipment. For instance, from the 24th century onwards this was the basis of food replication. Starfleet's transporters were used to analyze and record countless meals, which were then transported at a low resolution—molecular instead of quantum—but not reassembled. Instead, the relatively simple patterns are stored in the ship's computers and can be used to rematerialize copies of the original meal at the press of a button. This means that people can eat meat without having to kill any animals.

Replicator technology can also be used to create spare parts, medicines, and even parts of rooms for use in holodecks.

RISKY PROCEDURE

Although they are extremely useful and highly reliable, over the years a surprising number of things have gone wrong with transporters. The most obvious dangers encountered when using early models were that a pattern would degrade, killing the person in mid-transport, or that foreign objects would get mixed up in the matter stream and be combined with the person who was transported.

In early experiments beams weren't rematerialized in sufficient time and the person being transported was stuck in an incomplete matter stream. This happened to Cyrus Ramsay, one of the first test subjects for long-range transports, who failed to rematerialize after an experimental transport.

In 2209 it was also discovered that repeated journeys through primitive transporters could lead to a breakdown of certain neuro-chemicals, causing a loss of motor control and affecting higher brain functions. People who suffered from the condition, which was dubbed 'transporter psychosis', became paranoid, and suffered from hallucinations, painful spasms, and dementia. Once the condition was identified transporters were modified by the introduction of multiplexed pattern buffers and it was eliminated.

Transporters have also been involved in a number of freak accidents that have proved difficult to replicate. The most notable involve splitting the person being transported so that two copies materialize rather than one. The first recorded incident of this was in 2266 when Captain Kirk was duplicated when beaming up from the planet Alfa 117. The two versions were physically identical but somehow Kirk's personality traits had been split between the two, with one version retaining reason and compassion but lacking his aggressive and decisive characteristics, and the other having the exact opposite traits. The two Kirks were eventually reintegrated by putting them through the same transporter beam.

A similar accident occurred in 2361 when the U.S.S. Potemkin was evacuating a science outpost on Nervala IV. The planet has an unusual distortion field that interferes with the operation of the transporters, and when Lt William Riker beamed back to the Potemkin his pattern had to be reinforced with a second

00.998	01.027	01.105	01.132	01.190	01.204	01.216	01.221
Pattern buffer begins acceptance of matter stream	Verification of matter stream integrity	Increase phase transition coil input power to 37 MeV	ACB to 1.9 MeV. Reference beam phase lock	Reverify target coordinates, range, and relative velocity	Reverify integrity of pattern buffer operation with option to switch to backup buffer or abort sequence	Target lock. Begin continuous scan of target coordinates	Emitter array begins transmission of annular confinement beam to target coordinates

03.069	04.077	04.185	04.823	04.824	04.947	04.949	04.951
50% transmission of matter stream reached—option to abort cancelled	Dematerialization cycle completed. ACB power level maintained	Phase transition coils maintained at 25.1 Ghz	Emitter array materialization sequence complete	Verification of pattern integrity	Phase transition coils power down to standby	Primary energizing coils release ACB lock	Superconducting tokamak power down to standby

containment beam. Somehow the original beam was reflected back to the surface, causing Riker to rematerialize there even though the second beam successfully transported him on to the ship. Thus the transporter created a perfect clone of him. Unlike the duplicate Kirks, the Rikers were completely identical. At the time no one realized what had happened and one Riker went on to become first officer of the *Enterprise*-D, while the other remained on the planet until he was eventually discovered in 2369.

The reverse process where two living beings are combined into one has also been observed. In 2372 on board the *U.S.S. Voyager* the Vulcan Tuvok and the Talaxian Neelix emerged from a transport as a single living being, who became known as Tuvik. The procedure was ultimately reversed and both men were restored in full health, though this inevitably involved the destroying of Tuvik's life.

In other incidents people have been reassembled slightly out of phase with the normal universe, making them invisible and even intangible. This happened to Hoshi Sato in 2152 on one of *Enterprise* NX-01's early missions and to Lt Ro Laren and Lt Commander Geordi La Forge on the *Enterprise* NCC-1701-D after they interacted with an experimental Romulan cloaking device.

It has even been discovered that certain life forms can exist in the matter stream. The *Enterprise*-D crew detected quasi-energy microbes that had infected one of their

crewmembers in the transporter stream and existed as both matter and energy simultaneously. They were eventually removed by reprogramming the biofilters.

Transporters have also proved a way of accessing parallel universes. When a beam passes through an ion storm it can be shifted into the so-called Mirror Universe where the Federation—known there as the Terran Empire—is an aggressive military force that controls much of the Galaxy.

By the 2260s the transporter was the standard method for sending landing parties to a planet's surface. Personnel could be beamed down safely and quickly.

01.227	01.229	01.230	01.237	01.240	01.241	02.419	02.748
First detected return of ACB reflection. Doppler compensation syncronization with pattern buffer	Ground-level correction determination for target coordinates	ACB to full power	Pattern buffer begins transmission of matter stream to emitter array	Emitter array begins transmission of image data through ACB	Emitter array begins transmission of matter stream through ACB, and commences materialization sequence	Verification of materialization sequence. Option to abort and divert to alternate transporter pad	Phase transition coils begin ramp-down to 25.1 Ghz. [Commencement variable according to payload mass]

04.973	05.000	
Emitter array release ACB lock	Verification of successful transport	The times detailed cover the key events of the autosequence beamdown program for a standard Starfleet transporter. The specific times are variable depending on payload mass and transport range.

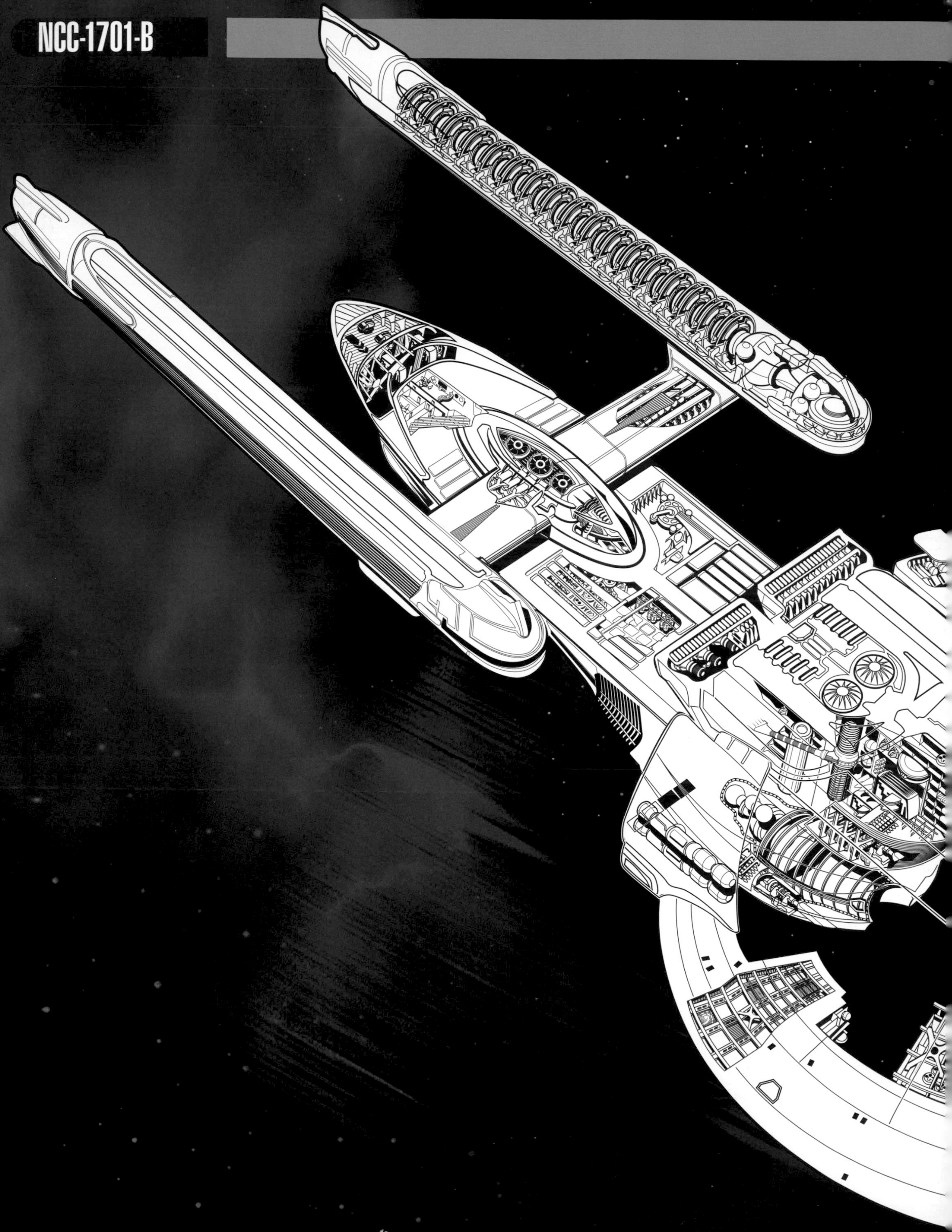

U.S.S. Enterprise
NCC-1701-B

The *U.S.S. Enterprise* NCC-701-B was an *Excelsior*-class ship that was built at Starfleet's Antares shipyards. She was formally commissioned in Earth Spacedock in 2293. She is probably most famous because Captain Kirk was lost on her maiden voyage, but she also undertook significant deep-space exploration missions and was involved in the military operations during the Tomed Incident, which led to the creation of an important treaty between the Federation and the Romulan Empire. She remained in service for 36 years before she was lost in deep space. Her exact fate has always been a mystery though it is likely that the crew fell victim to a plague.

Classification	Excelsior class [Enterprise variant]
Constructed	Antares Shipyards
Launch date	2293
Lost in action	2329 [fate unknown]
Length	467m
Number of decks	32
Crew complement	502
Weaponry	Phasers and photon torpedoes
Commanding Officers	John Harriman, William George, Demora Sulu, Thomas Johnson Jr

The *Enterprise*-B encountered the Nexus Ribbon when it responded to a distress call from two El-Aurian vessels. The rescue mission was successful but the visiting Captain Kirk was lost, presumed dead.

Work on the *Enterprise* NCC-1701-B began in 2288 as part of Starfleet's project to replace the ageing *Constitution*-class ships with larger and faster vessels. The project began in the early 2280s, but it was delayed by several years after difficulties with Starfleet's new transwarp engine design. The *U.S.S. Excelsior*, the first ship in the class, initially entered service in 2284, when she was used as the test bed for the experimental transwarp drive. However, despite early promise this technology proved unreliable and was abandoned in 2287. This necessitated a complete replacement of the engine systems with more conventional technology, and the *Excelsior* finally entered full service in 2290, four years behind schedule.

The *Enterprise*-B followed three years later and was formally commissioned three months after her predecessor—*U.S.S. Enterprise* NCC-1701-A—was retired. Her construction was supervised by Captain John Harriman, who had limited experience of active space exploration. Since the previous *Enterprise* had been so celebrated there was considerable interest in the launch of the new ship, and Captain James T. Kirk and several members of his senior staff attended the event. The plan was to make a brief journey to Pluto and back, but during the ceremony the ship detected two El-Aurian vessels in serious distress after they were caught in an energy ribbon known as the Nexus. The *Enterprise*-B responded and succeeded in rescuing one of the ships and 47 crewmembers from the other vessel, but Captain Kirk was lost during the mission while making the modifications to the ship's deflector dish that allowed her to escape from the ribbon.

The *Enterprise*-B was badly damaged during the mission and only left Spacedock several months later, this time under Captain William George. His senior staff included Chief Engineer Michael Jennings, Chief Medical Officer Dr Kate Giles, and science officer Narayn'Ka. During Captain George's command the ship was instrumental in exploring the area of space beyond the Gourami Sector, during which time she mapped 142 star systems and made first contact with 17 civilizations.

Following two tours of deep-space exploration, the *Enterprise*-B was assigned to Federation space, where part of her duties involved patrolling the border of the Romulan Neutral Zone. The 2290s had seen some diplomatic progress between the Federation and the Romulan Empire—for example, both parties were involved in productive talks at the Khitomer conference and at the negotiations that followed. However, in the early years of the 24th century tensions between the two powers began to grow again and the *Enterprise*-B was involved in stand-offs with Romulan warships on a number of occasions.

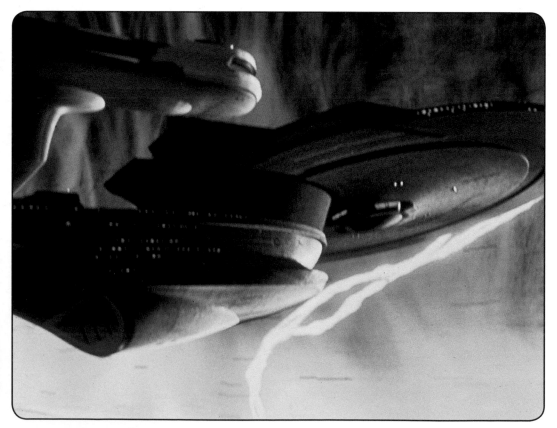

Events culminated in the disastrous Tomed Incident, in which Starfleet responded to a Romulan attack that cost thousands of Federation lives by fitting several of its ships, including the *Enterprise*-B, with cloaking devices and sending them into the Romulan Neutral Zone. The potential conflict caused a split in the Romulan Senate that was only resolved when an isolationist faction assumed power and agreed to begin peace negotiations with the Federation. The *Enterprise*-B transported several Federation ambassadors, including the Vulcan Sarek, to the planet Algeron, where a new treaty was negotiated that re-established the Neutral Zone and prohibited the Federation from using cloaking devices.

Following the Tomed Incident the *Enterprise*-B's first officer, Demora Sulu, who had joined the ship as helm officer straight from Starfleet Academy, was promoted to captain. During the following years she commanded the ship on a variety of voyages, including a two-year mission to chart the archaeological remains of the proto-Vulcan Debrune civilization, in the Barradas system.

In the 2320s *Enterprise*-B was reassigned to the newly established border with Cardassian space. Now under the command of Captain Thomas Johnson Jr, the *Enterprise*-B was again involved in a very tense

situation as the two powers came close to all-out conflict. When the Cardassian Union annexed the planet Bajor in 2328, *Enterprise*-B offered assistance to a number of Bajoran refugee ships, but Starfleet was unwilling to enter into a full-scale conflict and *Enterprise*'s role was restricted to relocating the Bajorans to nearby planets in Federation space.

The *Enterprise*-B was lost, presumed destroyed, in 2329. The last reports Starfleet received indicated that the crew had contracted a dangerous infection, but exactly what happened after the ship's final transmission is unknown.

The *Enterprise*-B was a modified *Excelsior*-class ship that entered service in 2293.

The construction of the *Enterprise*-B was supervised by Captain John Harriman, who had little experience of active service in deep space.

PORT ELEVATION

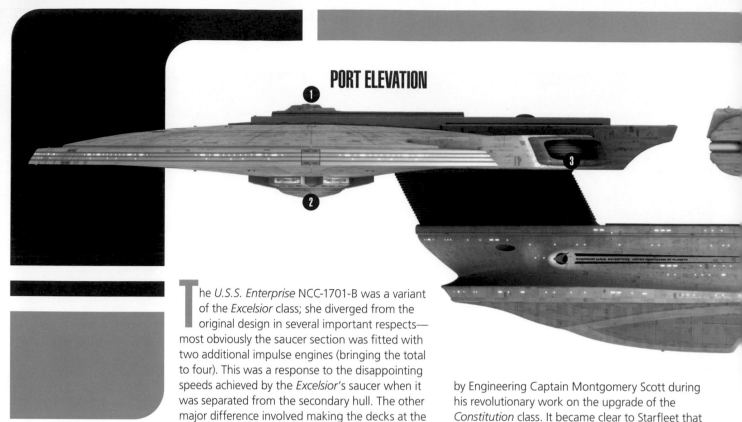

The *U.S.S. Enterprise* NCC-1701-B was a variant of the *Excelsior* class; she diverged from the original design in several important respects—most obviously the saucer section was fitted with two additional impulse engines (bringing the total to four). This was a response to the disappointing speeds achieved by the *Excelsior*'s saucer when it was separated from the secondary hull. The other major difference involved making the decks at the bottom of the engineering hull wider to provide additional lab space and sensor capacity.

Despite its troubled beginnings as the test bed for Starfleet's failed transwarp experiments of the 2280s, the *Excelsior* class was one of the Federation's most successful starship designs. The fundamental engineering principles behind this class of ship drew heavily on the innovations developed

by Engineering Captain Montgomery Scott during his revolutionary work on the upgrade of the *Constitution* class. It became clear to Starfleet that the ideas that were being developed could be better exploited in an entirely new class of ship and this led them to start work on the *Excelsior*. The final design proved incredibly flexible and remained in service for the next century.

Internally the *Enterprise*-B was a standard *Excelsior*-class ship. She had 32 decks, with the main bridge on Deck 1 at the top of the saucer

BOW ELEVATION

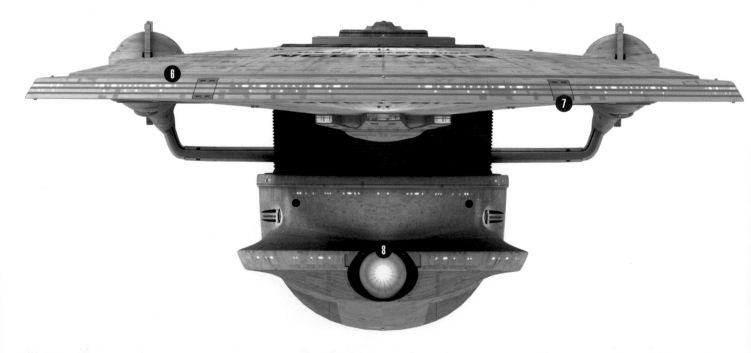

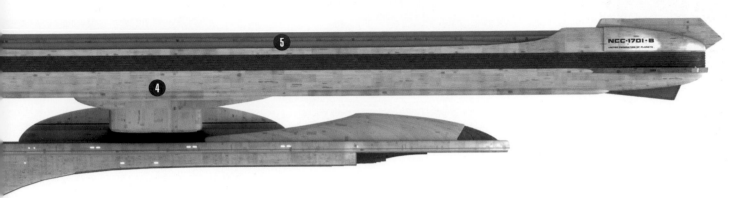

section, and Deck 32—where antimatter was both generated and stored—at the bottom of the secondary hull.

The saucer section provided almost all of the ship's living quarters, with the majority of the officers being billeted on Deck 4. The main sickbay was on Deck 6, at the dead center of the saucer, ensuring that it was one of the safest locations on the ship. In her original arrangement junior crewmembers had to share quarters, but she was subsequently reconfigured to provide each crewmember with more living space for deep-space exploration missions. This was considered necessary given the extended periods that the crew had to spend on board ship.

When she launched, the *Enterprise*-B was one of the fastest and most technologically advanced ships

in the fleet. She was only the fifth *Excelsior*-class ship to enter service.

In a design refinement carried over from the *Excelsior* prototype the warp core was positioned

1 Main bridge

2 Sensor dome

3 Impulse engine

4 Warp nacelle

5 Warp field grille

6 Phaser array

7 RCS thrusters

8 Navigational deflector

9 Subspace field radiator

10 Impulse engines

11 Main shuttlebay

12 Probe launch bay

STERN ELEVATION

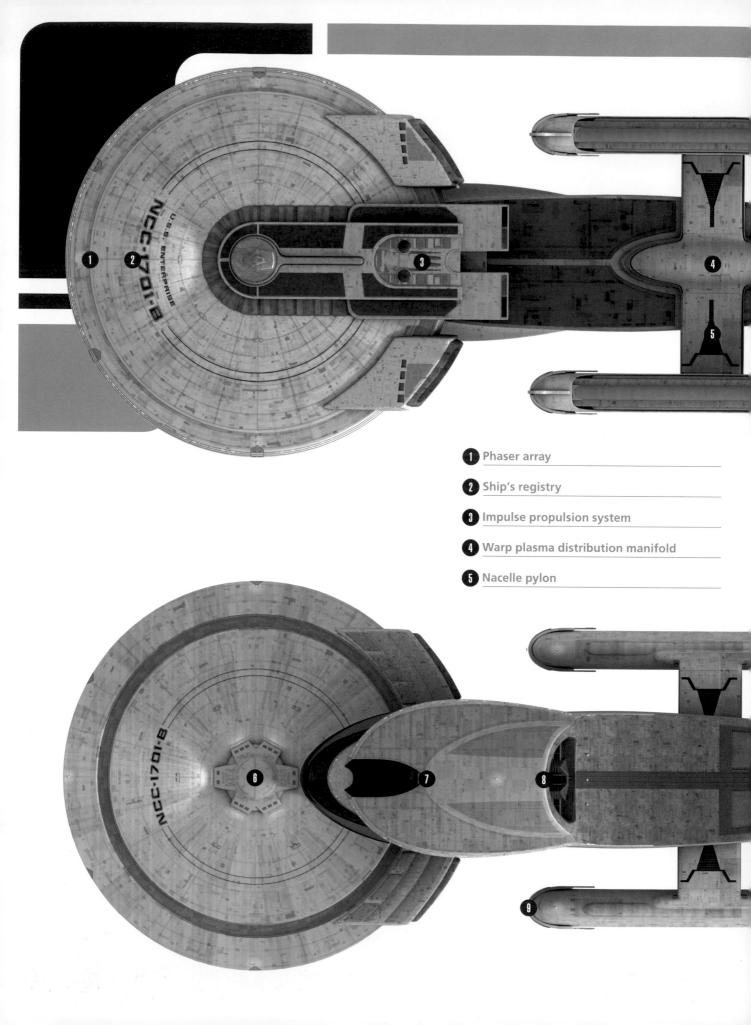

1 Phaser array

2 Ship's registry

3 Impulse propulsion system

4 Warp plasma distribution manifold

5 Nacelle pylon

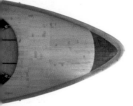

DORSAL ELEVATION

6 Sensor dome

7 Navigational deflector

8 Probe launch bay

9 Bussard ramscoop

10 Aft defensive systems

VENTRAL ELEVATION

TRANSWARP DRIVE

In the 2280s Starfleet fitted an experimental form of warp engine to the *U.S.S. Excelsior* NX-2000, which they referred to as a transwarp drive. It was hoped that the technology would be a major breakthrough, and the *Excelsior* conducted tests for the best part of a decade under the command of Captain Styles.

Transwarp drive relied on an extremely complicated set of equations that boosted the power of a conventional warp engine in much the same way that the Kelvans had modified the *Enterprise* NCC-1701's engines in 2268. However, although the system could be made to work in computer models, in practice it proved unworkable and the *Excelsior* never managed to achieve the kind of sustainable speeds the project predicted.

The transwarp project was formally abandoned in 2287 and all subsequent *Excelsior*-class vessels, including the *U.S.S. Enterprise* NCC-1701-B were fitted with conventional warp engines. Ironically, due to advances in conventional warp theory, the *Enterprise*-B eventually attained greater speeds than those projected for the transwarp project in the 2280s.

By the 24th century transwarp was used to describe any velocity that exceeded conventional warp speeds and thus could cover a variety of technologies. For example, it was used to describe both the Borg's network of subspace corridors and the Warp 10 flight that was made by Tom Paris on the Cochrane shuttlecraft.

toward the front of the ship and ran almost the entire height of the engineering hull from Decks 12 to 31.

There were two computer cores, one in each hull. If the saucer was separated they could operate independently from one another, but in normal operations the duotronic systems were in constant communication, with each providing a backup for the other in the case of a significant failure, such as the one the *Excelsior* experienced in 2285. The computer core in the saucer section ran between Decks 6 and 9. Its larger counterpart in the secondary hull ran the height of nearly five decks, between Decks 23 and 28.

The *Enterprise*-B had a single shuttlebay at the rear of the secondary hull. The majority of the ship's cargo was brought in through here before being moved to the twin cargo bays on the top of the secondary hull.

The *Enterprise*-B was a heavily armed vessel that was designed for combat as much as exploration. She was fitted with Type-8 phaser emitters and fore and aft photon-torpedo launchers. The large launch bay located in the secondary hull was also used to launch probes and subspace relays.

Time travel

SPACE-TIME MANIPULATION THEORY

For a long time it was assumed that time travel, as most people understand it, was impossible. Not only did it seem to break the fundamental laws of physics, but, as many people pointed out, if time travel were possible, surely we would be besieged by time travelers from the future? As far as the Vulcans were concerned, this last argument was *prima facie* proof that time travel was impossible.

However, even in Earth's early 20th century it was clear that time did not always progress at the same apparent rate. For example, Einstein produced a theoretical proof that the stronger gravity is, the more slowly time passes, at least relative to a place with less gravity. There appear to be many bizarre phenomena relating to time and Starfleet has encountered areas of space where time moves

SLINGSHOT MANEUVER

The first practical method of time travel developed by Starfleet involved flying a vessel at high speed into the gravitational field of a star. The ship has to fly dangerously close to the star, performing a half-orbit. When it breaks free of the strongest part of the star's gravity field in a 'slingshot' it causes a distortion in the space-time continuum, throwing the ship into the past or the future.

at different speeds, repeats itself, and even goes backwards.

There is also highly contested evidence that time travelers have been interfering with the normal passage of history. However, final proof that time travel is possible was only discovered in 2266 by the crew of the *U.S.S. Enterprise* NCC-1701 when they were forced to perform a cold start of their warp engines. Somehow the warp field generated a serious distortion in the fabric of space-time, causing the ship to travel 71 hours into the past. A few months later the *Enterprise* crew traveled much further into the past after an encounter with a black hole. This time they were stranded in 1969 and narrowly avoided damaging history. After the *Enterprise* was spotted by the US Air Force the crew took a pilot called John Christopher on board the ship. They considered taking him with them but realized that this would have prevented the birth of Colonel Shaun Christopher, who would command the first Earth–Saturn probe mission. Ultimately they were able to return

John Christopher to a moment in time just before he encountered them.

In order to return to their own time the crew developed a working theory of time travel. This involved combining the distortions in space-time caused by a starship's warp drive with the gravitational forces close to a star. The starship has to fly incredibly close to the star, breaking free at a carefully chosen moment, opening a channel through time. Determining the exact amount of time covered requires some extremely precise calculations involving the mass of the ship, the gravitational field of the star, and the distortion caused by the warp engines. However, it is a method of time travel that has been replicated successfully, not least in 2286 when the then Admiral Kirk and his crew used it to bring two humpback whales to the 23rd century.

It has also been shown that the fabric of space-time can be damaged by warp core breaches and substantial matter/antimatter detonations caused by repeated photon torpedo

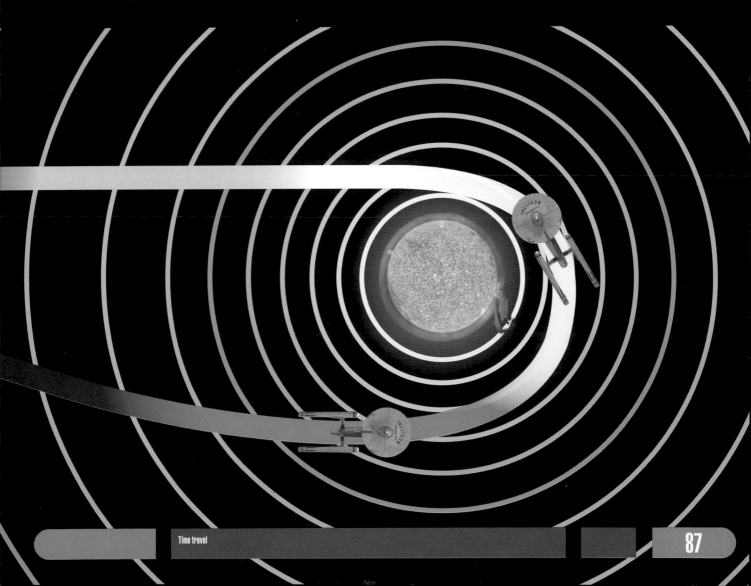

fire. In some cases these have been shown to open rifts through time. The *U.S.S. Enterprise* NCC-1701-C traveled through a rift created in this way during the battle at Narendra III. The results of this kind of damage are impossible to predict and it is not a reliable method of time travel.

Starfleet has discovered several more reliable methods of traveling through time. The most impressive is the Guardian of Forever, an extraordinary, sentient device that can create portals to anywhere in time and space. The Guardian was created over five billion years ago by an ancient and incredibly advanced race. The crew of the *U.S.S. Enterprise* NCC-1701 encountered it in 2267 and it was one of Captain Kirk's most significant discoveries during his initial five-year mission. However, the dangers of using it soon became apparent when on the first trip through it Dr McCoy saved the life of Edith Keeler, a charity worker in New York in the 1930s. In the new timeline she

campaigned for peace and this changed the course of history, delaying America's entry into World War 2 and allowing the Nazis to win. Kirk was forced to correct the timeline by allowing Keeler to die.

The *Enterprise* NCC-1701 crew also encountered time-portal technology on the planet Sarpeidon, shortly before it was destroyed by the supernova of the star Beta Niobe in 2269. It is not known how the Sarpeidon time-travel devices worked, but the inhabitants used them to create portals that allowed them to escape into their own past.

By the 24th century Starfleet had discovered chronitons, quantum particles that could be manipulated to distort space-time to allow time travel. These were used by the Borg to travel to Earth's past in an attempt to prevent Zefram Cochrane's first warp flight.

The possibility of time travel and the dangers it posed led the Federation to establish a Temporal Prime Directive, which instructed any personnel traveling through time to make minimal contact with people in the past, never to offer them assistance, and to keep any interference to an absolute minimum. The Federation Department of Temporal Investigations was set up to investigate any incidents of time travel. Perhaps unsurprisingly, James T. Kirk had the largest file in their records, with 17 separate violations.

The Federation has also encountered dimensions and even beings that exist outside normal space-time. Captain Kirk was swept into the Nexus, a dimension that exists outside time, in 2293, but was able to leave it in 2371 without ageing. The Bajoran Prophets, a race of incredibly powerful beings that live inside the Bajoran Wormhole, exist outside linear time. An artifact known as the Orb of Time, which apparently originated inside the wormhole, is also capable of sending people through time.

In the Delta Quadrant, the Krenim used time travel technology to launch a temporal weapon ship that existed outside normal space-time and could wipe their enemies out of history. Unfortunately for them, the consequences of their interference with the timeline were too difficult to predict and they kept altering history in a desperate attempt to restore everything they had destroyed. Ultimately the weapon ship itself was erased from history, restoring the original timeline before the Krenim began to interfere with its course.

The Guardian of Forever is one of the most extraordinary methods of time travel Starfleet has ever encountered. It appears to be a sentient portal that can send a person to any point in space and time.

Admiral Kirk and his crew used a stolen Klingon ship to visit San Francisco in the 1980s so that they could repopulate Earth's oceans with humpback whales.

Starfleet is also familiar with the Q, an apparently omnipotent race, who can travel through time using nothing more than thought.

In the future it seems that time travel will become commonplace but no less dangerous. It appears that by the 31st century there will be a temporal cold war, with different factions interfering in the past for their own benefit. The original *Enterprise* NX-01 was one of the focal points of this temporal cold war. Captain Archer seems to have prevented any serious damage, but, of course, the difficulty with time travel is that no one can ever be certain.

The crew of the *Enterprise*-D found themselves taking part in Zefram Cochrane's historic first warp flight when they pursued the Borg into Earth's past so they could protect the history of humankind.

Captain Archer found himself involved in a temporal cold war in which various factions from the future tried to interfere with history.

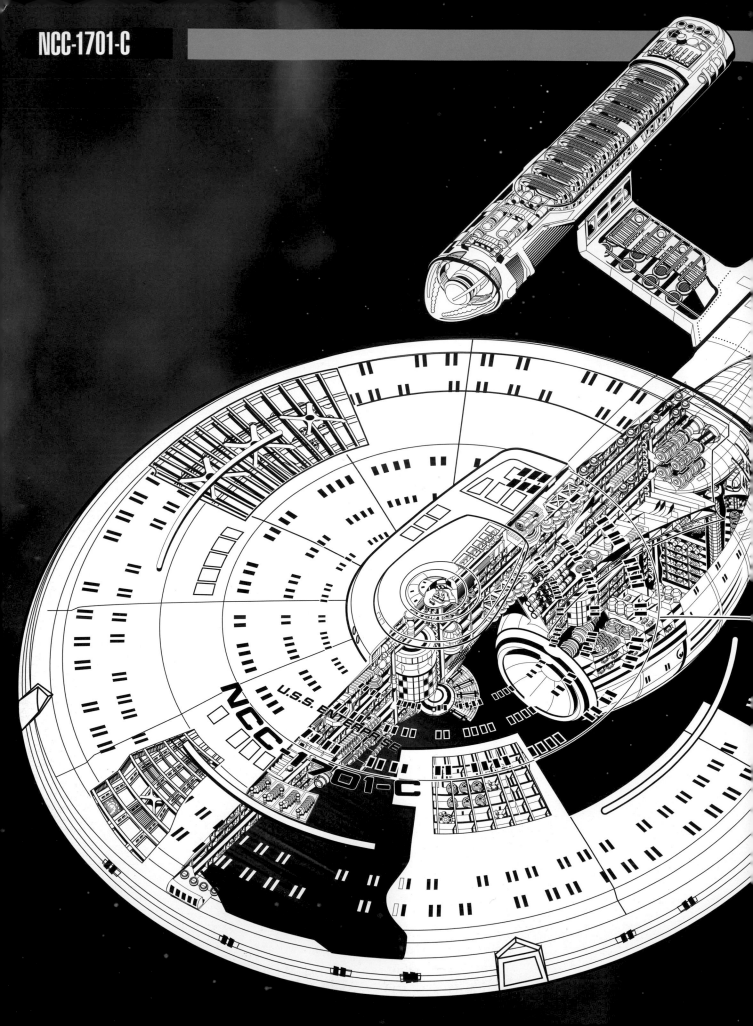

U.S.S. Enterprise
NCC-1701-C

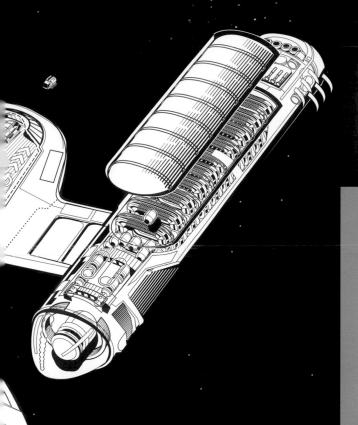

The fourth Federation vessel named *Enterprise* was an *Ambassador*-class ship that was launched in 2332, under the command of Captain Rachel Garrett. The *U.S.S. Enterprise* NCC-1701-C served with great distinction for 12 years before she was destroyed in 2344 after coming to the defense of a Klingon outpost at Narendra III, which was being attacked by four Romulan Warbirds. The sacrifice of the *Enterprise*-C played an important role in cementing relations between the Federation and the Klingon Empire. The Klingons place great value on bravery and the crew's willingness to sacrifice their ship to defend a small Klingon outpost was greatly admired by the Klingon High Council, who saw it as evidence that the Federation could be trusted.

Classification	Ambassador class
Constructed	Earth Station McKinley
Launch date	2332
Destroyed	2344 [defending the Klingon outpost at Narendra III]
Length	526m
Number of decks	36
Crew complement	530
Weaponry	Phasers and photon torpedoes
Commanding Officers	Rachel Garrett

AMBASSADOR CLASS

LIST OF VESSELS

The *Enterprise* NCC-1701-C was the third *Ambassador*-class vessel to be built and was commissioned in 2332 at Earth Station McKinley. Command was given to 33-year-old Captain Rachel Garrett, who was promoted after an impressive tour of duty as the first officer of the *U.S.S. Hood*.

The 2330s were a tense period. Although the Federation had been in peace negotiations with the Klingon Empire since the 2290s, relations between the two powers were often strained. From 2341 onwards *Enterprise*-C was assigned to an area of space that bordered both the Klingon and Romulan Empires. The Federation had had no direct contact with the Romulans since the Tomed Incident in 2311, when thousands of lives were lost. However, Starfleet was aware that the Romulans and Klingons were at odds and the Romulans mounted assaults on Klingon colonies throughout the 2340s.

In 2344 the *Enterprise*-C was on course for the planet Archer IV when she responded to a distress call from a Klingon outpost on Narendra III. When she arrived Captain Garrett discovered that the small Klingon settlement was facing four Romulan Warbirds and had no chance of survival. She attacked the Romulans even though her ship was heavily outgunned. During the battle, the *Enterprise*-C took more than 400 casualties and lost her entire senior staff apart from Captain Garrett. At the decisive moment in the battle a photon torpedo created a rift in space-time and the *Enterprise*-C was thrown 22 years into the future.

In the future that the *Enterprise*-C visited, the Federation was at war with the Klingon Empire following the breakdown of peace negotiations in the 2350s. Garrett's *Enterprise* encountered her immediate successor, the *Galaxy*-class *U.S.S. Enterprise* NCC-1701-D under the command of Captain Jean Luc Picard. He decided that the *Enterprise*-C's departure from the timeline might well have had serious consequences and persuaded Captain Garrett to return to her own

The *Enterprise*-C visited a version of 2366 in which the Federation was losing a war with the Klingons. All Starfleet's ships, including the *Enterprise*-D, had become military vessels.

time even though she and her crew were facing certain death.

While the *Enterprise*-C was in the future she was repaired and made battle-ready; unfortunately, she was also attacked by the Klingons and Captain Garrett was killed, leaving Lt Richard Castillo to take command. He was assisted by Lt Tasha Yar, a member of the *Enterprise*-D crew, who decided to join him on the *Enterprise*-C.

When Castillo returned to Narendra III only seconds had passed in normal time and the ship was able to inflict serious damage on the Romulans, destroying one of the Warbirds before she was eventually overpowered. The surviving 36 members of the crew, including Tasha Yar, were taken prisoner by the Romulans and transported to the Romulan home world. The Federation only learned of their capture in 2368.

However, the *Enterprise*-C's sacrifice did not go unnoticed. The Klingons detected the remains of the destroyed Warbird and identified the weapons signature as belonging to a Starfleet ship. They greatly admired the sacrifice of the *Enterprise* and regarded Captain Garrett's willingness to enter a battle she stood little chance of winning as both honorable and glorious. More importantly, they saw it as proof that the Federation could be a worthy ally. The heroic actions of Captain Garrett and her crew also ensured that the future that they visited never came to pass; instead, the Federation and the Klingon Empire became allies, ushering in a much more peaceful period of history.

Captain Garrett survived the journey to the future, but was shocked to learn that history required her crew to sacrifice themselves for the greater good.

The *Enterprise*-C's battle with the Romulan ships created a distortion in the space-time continuum that led to a possible future.

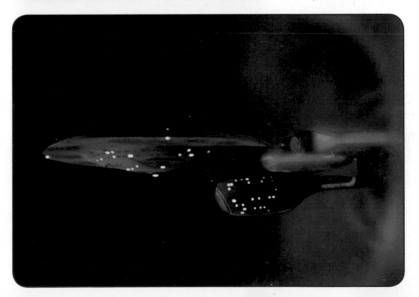

PORT ELEVATION

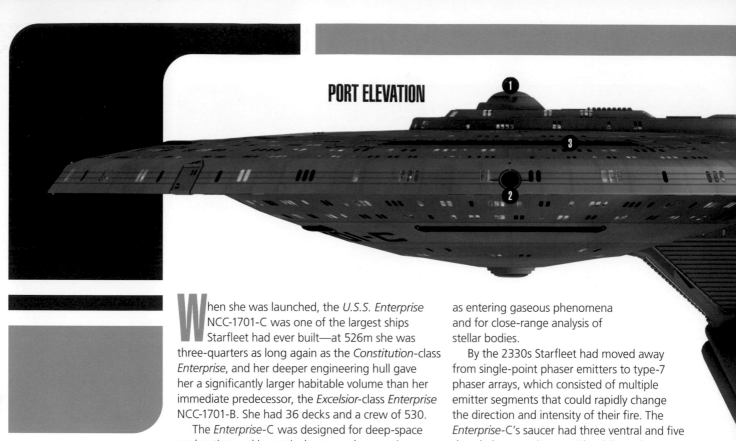

When she was launched, the *U.S.S. Enterprise* NCC-1701-C was one of the largest ships Starfleet had ever built—at 526m she was three-quarters as long again as the *Constitution*-class *Enterprise*, and her deeper engineering hull gave her a significantly larger habitable volume than her immediate predecessor, the *Excelsior*-class *Enterprise* NCC-1701-B. She had 36 decks and a crew of 530.

The *Enterprise*-C was designed for deep-space exploration and in particular to catalog a variety of stellar phenomena, and accordingly she was equipped with an unusually high number of shuttles. *Ambassador*-class ships had two shuttlebays—as was traditional, one was located behind clamshell doors at the rear of the engineering hull, but this class of ship also had a further shuttlebay in the saucer section. The shuttles were designed to perform missions where transporters were ineffective, such

as entering gaseous phenomena and for close-range analysis of stellar bodies.

By the 2330s Starfleet had moved away from single-point phaser emitters to type-7 phaser arrays, which consisted of multiple emitter segments that could rapidly change the direction and intensity of their fire. The *Enterprise*-C's saucer had three ventral and five dorsal phaser emitters, with additional arrays on the lower edges of the nacelle pylons and at the rear of the secondary hull under the shuttlebay.

The *Enterprise*-C also had fore and aft photon-torpedo launchers on the neck that connected the secondary and primary hulls.

The design of the warp engine and the field coils incorporated design elements from the failed transwarp experiments of the 2280s. The impulse

BOW ELEVATION

SYSTEMS OVERVIEW

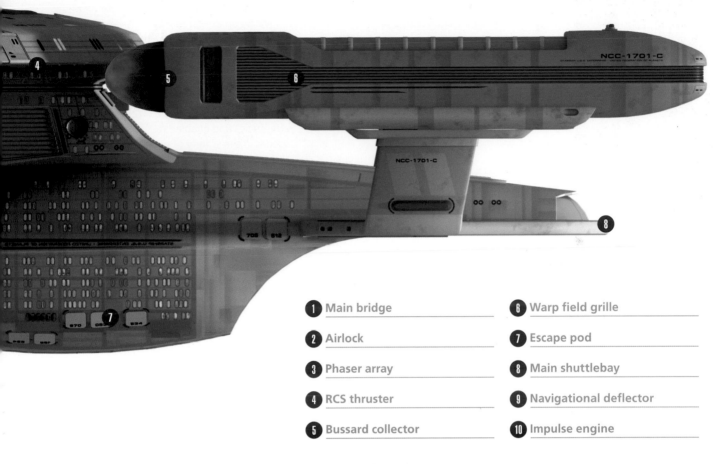

❶ Main bridge	❻ Warp field grille
❷ Airlock	❼ Escape pod
❸ Phaser array	❽ Main shuttlebay
❹ RCS thruster	❾ Navigational deflector
❺ Bussard collector	❿ Impulse engine

STERN ELEVATION

engines also featured a significant upgrade from the previous generation of vessels and the *Enterprise*-C was one of the most maneuverable ships of her era.

The *Ambassador* class was one of the first Starfleet designs that could eject its warp core in case of a catastrophic matter/antimatter containment failure. The warp core ran the height of the secondary hull with a hatch in the under side. In an emergency the chief engineer could blow the hatch with explosive charges, disconnect the warp core from the matter and antimatter injectors, and eject it.

The computer systems were among the first to use isolinear circuitry, an advanced optical system that replaced the enhanced duotronic circuitry that had been in use on starships since the 2240s.

The pylon design reflected changes in the Advanced Starship Design Bureau's (ASDB) thinking about warp field generation. As on later ships, the nacelles were dropped to a position slightly below the saucer section. This design would continue with the *Galaxy* and *Intrepid* classes. The altered geometry of the warp field produced significant efficiencies and enabled the ship to sustain cruising speeds in the vicinity of warp 8.4.

1 Ship's registry

2 Main bridge

3 Escape pod

4 RCS thruster

5 Shuttlebays

6 Phaser array

7 Navigational deflector

8 Warp core ejection hatch

DORSAL ELEVATION

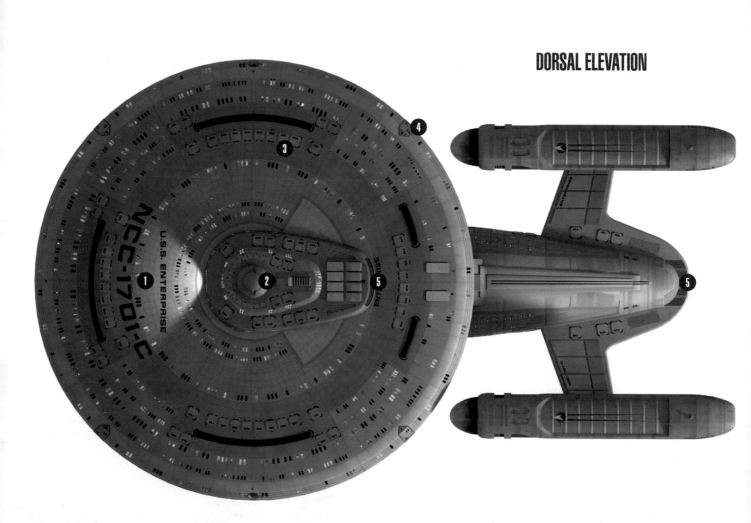

WARP FIELD DEVELOPMENT THEORY

The warp "bubble" in which a starship rides at faster-than-light speeds is not a static structure. Rather, it is a flowing, pulsing, ever-shifting field of energy that dictates the shape of a particular starship. Even minor changes in starship design must undergo extensive testing and optimization to ensure safety and propulsive efficiency.

Field geometry is particularly sensitive to variations in the position of the warp engine nacelles. While in conventional (slower-than-light) spacecraft design, the propulsive thrust vector must travel through the vehicle's center of mass, warp propulsion is quite different. Warp drive requires the nacelle axes to be offset from the vehicle's centerline to create the propulsive imbalance. This technique usually requires the forward field lobe (normally containing the saucer section) also to be offset from the nacelle vector. In many older ship designs the nacelles were located slightly above the forward lobe. The *NX* and *Constitution*-class starships used this design, which yielded superior field stability, although it also resulted in reduced propulsive efficiency.

During the 2260s, the Advanced Starship Design Bureau began to experiment with field geometries that lowered the nacelle vector with respect to the forward field lobe. This approach promised improved engine performance but required several years of research into new field-stability software. Among the first vehicles to employ the new approach was the experimental Starship *Excelsior*, whose design lowered the nacelle centerline so it was level with the saucer module. This design was reflected in the *Excelsior*-class *Enterprise*-B. Further advances in field control software allowed even more radical designs for the *Ambassador*-class *Enterprise*-C, and the *Galaxy*-class *Enterprise*-D, both of which saw the nacelles move even lower than the forward lobe.

Strangely enough, recent advances in deflector shield technology have resulted in a reversal of this trend in nacelle position in some of the latest ships. New ultra high frequency shield generators have allowed the *Sovereign*-class *Enterprise*-E to employ higher nacelle positions, while enjoying an additional 3 per cent in engine performance.

The arcane engineering of warp field geometry continues to be a subject of intense study at Starfleet's Advanced Starship Design Bureau and will undoubtedly continue to be an area of ever-changing research and innovation.

VENTRAL ELEVATION

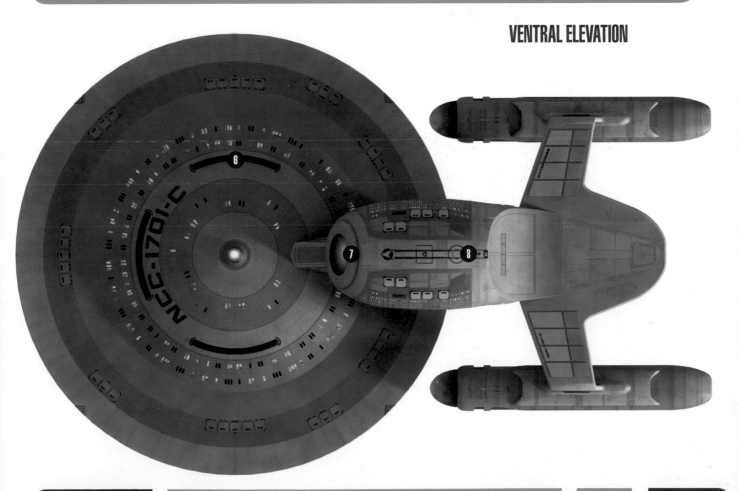

Deflector shields
HULL BREACH PREVENTION THEORY

One of the dangers of traveling at huge speeds is that if you collide with something, however small, it will produce an enormous impact that could easily puncture the hull of a starship. And, since starships have to operate in the vacuum of space, it's vitally important that they stay airtight. The threats posed to starships range from the obvious—such as small asteroids—to micrometeoroids, particles of dust, or even stray hydrogen atoms.

Ships are obviously designed with reinforced hulls that make them tough enough to resist minor impacts. In fact, early shuttles were completely dependent on hull polarization. However, the only real solution is to avoid collisions in the first place.

The navigational systems are designed to avoid colliding with large hazards such as

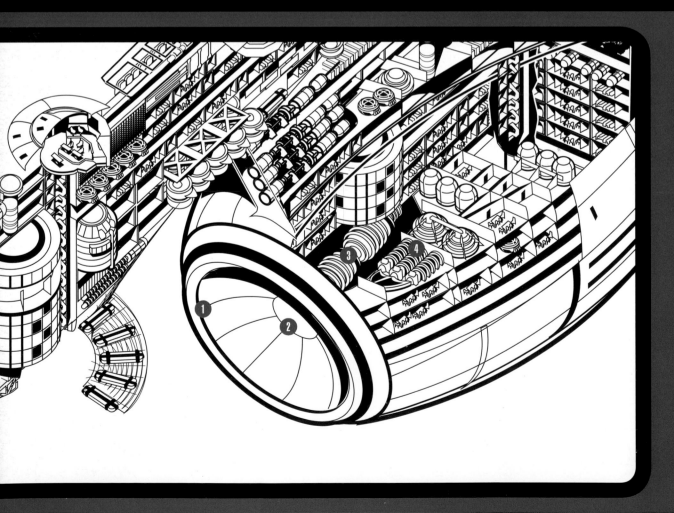

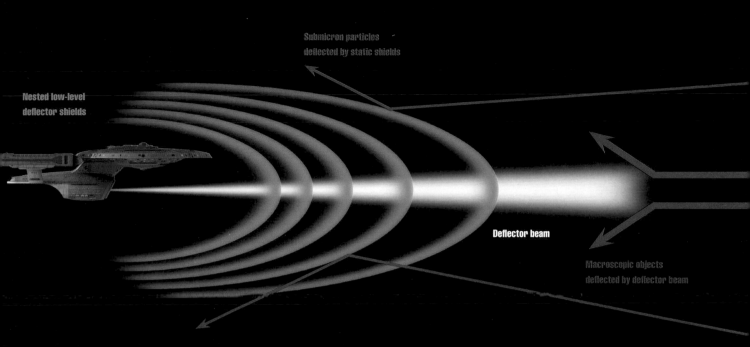

Submicron particles
deflected by static shields

Nested low-level
deflector shields

Deflector beam

Macroscopic objects
deflected by deflector beam

planets and asteroids. The on-board computers are programmed with all the available data about the position and speed of movement of stars, planets, and stellar phenomena such as comets. This is known as the Galactic Condition Database and is gathered from a variety of sources including deep-space telescopes such as the Argus Array, and is regularly updated with reports from starships. The ship's on-board sensors supplement this data with immediate up-to-date readings, and the on-board navigational computers make automatic course corrections to avoid colliding with anything.

However, this doesn't take account of the serious threats posed by smaller objects. This is the role of the navigational deflector. On every version of the *Enterprise* this has been a beam emitter near the front of the ship, and on every model except the NX-01—where it was built into the front of the saucer—it has been at the front of the engineering hull.

The navigational deflector emits a deflector beam that sweeps thousands of kilometers ahead of the ship, clearing everything away from its path. On later starships such as the *Enterprise*-D the deflector actually emits two different sets of fields. The first of these is an active beam that sweeps a path several thousand kilometers in front of the ship. This beam, which is essentially an inverted tractor beam, uses data from the long-range sensors to detect relatively small objects such as meteoroids, which it then pushes out of the

ship's path. The deflector can be adjusted to alter the direction of the beam.

The deflector also generates a series of nested deflector fields directly in front of the ship. These relatively low-level fields cover a distance of roughly 2km and deflect microscopic particles before they strike the ship.

The *Enterprises* that were designed to separate their saucer and engineering sections, but still operate at high speeds, have a secondary deflector that can be found on the leading edge of the saucer. This also provides backup if the main deflector fails.

In an emergency the deflector beam can also be used as a weapon. It was designed to channel an enormous amount of power—far greater amounts than the phaser arrays. If enough power is directed through it the deflector can produce a devastating energy beam, but since it was not designed for this purpose it burns out quickly. In practice a deflector can only be used this way once before it is completely rebuilt.

1 Parabolic reflector

2 Multiphase emitter matrix

3 Primary beam focusing coils

4 Secondary power coils

U.S.S. Enterprise
NCC-1701-D

The *U.S.S. Enterprise* NCC-1701-D was the sixth Starfleet vessel to carry the illustrious name and proudly continued the tradition of going boldly where no one had gone before. Commanded by Captain Jean-Luc Picard, the *Enterprise*-D was launched in 2363 and served for just eight years as the flagship of Starfleet before she was destroyed during an operation to prevent the destruction of the Veridian System in 2371.

Despite her brief operational life, the *Enterprise*-D had an extremely eventful existence and made more of an impact on the history books than any Starfleet vessel since Captain Kirk's *Enterprise* NCC-1701 of more than a century earlier.

Classification	Galaxy class
Constructed	Utopia Planitia Fleet Yards
Launch date	2363
Destroyed	2371 [warp-core breach under Klingon attack]
Length	641m
Number of decks	42
Crew complement	1,014 [typical]
Weaponry	Phasers and photon torpedoes
Commanding Officers	Jean-Luc Picard, Edward Jellico

GALAXY CLASS

The *U.S.S. Enterprise* NCC-1701-D was Starfleet's flagship and was in operation for just eight years before her untimely destruction in 2371. During that time her crew made first contact with 27 species, made countless scientific and cultural discoveries, engaged in several important diplomatic missions that prevented catastrophic wars, saved Earth and much of the Federation from Borg assimilation, and was nearly destroyed several times.

The *Enterprise*-D was 'home' to more than 1,000 crewmembers, and when she was launched in 2364 her senior crew comprised Captain Jean-Luc Picard, Commander William T. Riker as first officer, Commander Beverly Crusher as chief medical officer, Lt Commander Data as operations officer, Lt Commander Sarah MacDougal as chief engineer, Lt Commander Deanna Troi as ship's counselor, Lt Natasha Yar as tactical officer and security chief, and Lt Junior Grade Geordi La Forge as the conn officer. Other notable crewmembers were Lt Junior Grade Worf, initially also conn officer, Ensign Wesley Crusher, and Chief Miles

O'Brien as the transporter chief. Worf's assignment to the crew was particularly notable as he was the first Klingon to serve in Starfleet. He was promoted to lieutenant and became tactical officer following the death of Tasha Yar in late 2364.

In her first year of operation, the *Enterprise*-D had several different chief engineers. The first was Lt Commander Sarah MacDougal, followed by Lt Commander Argyle, Lt Logan, and Lt Commander Leland T. Lynch, before Geordi La Forge was promoted to full lieutenant and took on the role in 2365—a position he then held for the rest of the *Enterprise*-D's operational life.

Other notable changes to the crew included Commander Katherine Pulaski taking over as chief medical officer in 2365 for approximately a year, and the addition of Ensign Ro Laren as conn officer from 2368 to 2370.

During *Enterprise*-D's very first mission, she made the first of what would be several memorable encounters with Q, an omniscient, omnipotent, and obnoxious entity. Q put humanity on trial, accusing them of being a 'dangerous, savage child-race', and threatened them with extinction unless they could prove otherwise. Fortunately, Q agreed that humanity had some redeeming qualities and the potential to improve after the crew saved an enslaved and dying spaceborne entity that had been forced by the Bandi to take on the form of Farpoint Station. Also during the course of this mission, the *Enterprise*-D performed the first ever high-warp-speed saucer separation.

The *Enterprise*-D made first contact with several more spaceborne life forms during her missions, including Nagilum—an extremely powerful extra-dimensional creature that lived in a hole in space; Gomtuu—a living spaceship; the Crystalline Entity—a massive creature that resembled a snowflake and stripped planets of all organic life in order to sustain itself; and 'Junior'—a creature that 'fed' directly off the *Enterprise*'s power after its mother had accidentally been killed.

Other unusual life forms discovered by the *Enterprise*-D included microbrains—silicon-based life forms that resembled tiny sparkling crystals and referred to humans as 'ugly bags of mostly water'; Marijne VII beings—subspace life forms that resembled small chemical flames; and solanagen-based entities that existed in a deep subspace domain and abducted *Enterprise* crewmembers in their sleep to perform medical experiments on them.

The *Enterprise*-D even helped to create some new sentient life forms in the shape of nanites—submicroscopic robots that started off as part of an experiment by Wesley Crusher before becoming self-aware. Exocomps, engineering tools with artificial intelligence, also came to be regarded as sentient life forms while aboard the *Enterprise*-D when they demonstrated the instinct for self-preservation. A holodeck program that recreated the fictional character Moriarty was so sophisticated that it, too, was considered a sentient life form. Finally, the *Enterprise*-D herself became an emergent life form when the computer systems linked together, forming a neural network akin to a brain. This emergent life form manifested itself through existing characters in the holodeck, and as it became more complex it created semi-organic life forms that eventually left the ship to live in space.

As well as encountering new life forms, the *Enterprise*-D explored new areas of space. Thanks to a mysterious alien known as The Traveler, who had the ability to alter space and time merely with his thoughts, the *Enterprise* was hurled 2,700,000 light years to the distant M-33 galaxy. In trying to return her, The Traveler sent the ship a billion light years in the other direction before collapsing with exhaustion. Only when The Traveler had recovered did he manage to return the *Enterprise* to her home Galaxy.

In 2367 the *Enterprise*-D was sent by Q to another area of space that no Starfleet vessel had been to before—the Delta Quadrant. Here, she had her first encounter with the Borg, an immensely powerful society of cybernetic humanoids with one collective mind, who sought perfection by relentlessly conquering worlds and assimilating their technology. During this encounter the crew learned that the Borg could not be reasoned or bargained with and they heard for the first time the Borg's chilling mantra, 'Resistance is futile.' It soon became apparent that the *Enterprise*-D's tactical abilities were no match for those of the Borg's cube-shaped ship. The *Enterprise* was held in the Borg cube's tractor beam while a cutting beam sliced out sections 27, 28, and 29 on Decks 4, 5, and 6, killing 18 crewmembers in the process. Only the intervention of Q, who sent the *Enterprise* back to Federation space, saved the ship from certain destruction at the hands of the Borg, but Captain Picard felt that they had not seen the last of them—and he was right.

The following year a Borg cube invaded Federation space, destroying all before it as it headed straight for Earth. Picard was captured and assimilated, meaning that everything he knew about Starfleet defenses, the Borg now knew too. Commander Riker took over as captain of the *Enterprise*-D, which became part of a

Opposite: On Picard's first mission the *Enterprise*-D encountered Q, a seemingly omnipotent being who put all mankind on trial.

Above: The *Enterprise*-D was placed under the command of Captain Jean-Luc Picard, a veteran officer who had distinguished himself as captain of the *U.S.S. Stargazer*.

Below: The Traveler used his abilities to accelerate the *Enterprise*-D to unheard of speeds. He also revealed that Wesley Crusher had the potential to develop the same abilities.

40-strong armada of Starfleet ships assembled at Wolf 359, hoping to stop the Borg. Every ship except the *Enterprise*-D was destroyed in the confrontation, with the loss of 11,000 lives. The *Enterprise* was herself badly damaged, as Decks 23, 24, 25, and 36 were sliced open, with the loss of at least 11 lives. Her deflector dish and warp core were also overloaded as several decks were flooded with radiation, but the Borg cube was finally defeated when a computer command was successfully downloaded into the Borg Collective, ultimately causing its destruction.

In the aftermath of the battle with the Borg, the *Enterprise*-D required a full refit at Earth Station McKinley, which took six weeks. During the refit, the opportunity was taken to give her a phaser upgrade and a new dilithium chamber hatch, which later proved to be defective. This caused an explosion in the warp-drive system, which crippled the ship for two weeks.

For a vessel whose primary mission was to peacefully explore the Galaxy, the *Enterprise*-D had a surprising number of encounters that

almost led to her destruction. In 2368 she struck a quantum filament, a type of spatial anomaly, which caused widespread damage to the ship's systems, and several crewmembers were killed when they became trapped after emergency bulkheads closed. In addition the ship almost lost antimatter containment, which would have resulted in its destruction, but the crew managed to restore power before this happened.

Later the same year the *Enterprise*-D was destroyed in a collision with the *U.S.S. Bozeman*— over and over again—after she became trapped in a temporal causality loop. Fortunately, feelings of déjà vu among the crew allowed them to piece together the fact that they were living the same time period again and again and they were able to send a message into the next loop and thereby avoid the collision.

In 2369 the *Enterprise*-D was nearly destroyed in an explosion caused by a feedback loop when transferring energy to a Romulan Warbird, whose artificial quantum singularity warp core had failed. The destruction of both ships was avoided after it was discovered that life forms from another space-time continuum had colonized the artificial quantum singularity and fired on the *Enterprise* to protect themselves.

When the ship wasn't being nearly destroyed, or discovering new life forms, it was at the heart of political power struggles in the Alpha Quadrant. In 2368 the *Enterprise*-D played a central role in coordinating a tachyon detection grid that prevented cloaked Romulan ships from crossing the Klingon border and delivering supplies to the House of Duras during the Klingon Civil War. If the Romulans had been successful in helping Duras become the leader of the Klingon Empire, it would have led to a massive shift in the power balance of the Alpha Quadrant and almost certainly to war with the Federation.

In 2369 Captain Picard was sent on a clandestine mission to search for a Cardassian weapons research facility on Celtris III. While Picard was absent, Captain Edward Jellico took command of the *Enterprise*-D and prevented a Cardassian strike near the McAllister C-5 Nebula while also securing the release of Picard, who had been captured by the Cardassians.

The *Enterprise*-D was also instrumental in many scientific and cultural discoveries, the most important occurring in 2369 when evidence was unearthed that the Milky Way had been seeded by an ancient humanoid species. This proved that many species, including humans, Klingons, Romulans, and Cardassians, shared a common ancestor.

In 2370 the *Enterprise*-D became one of the

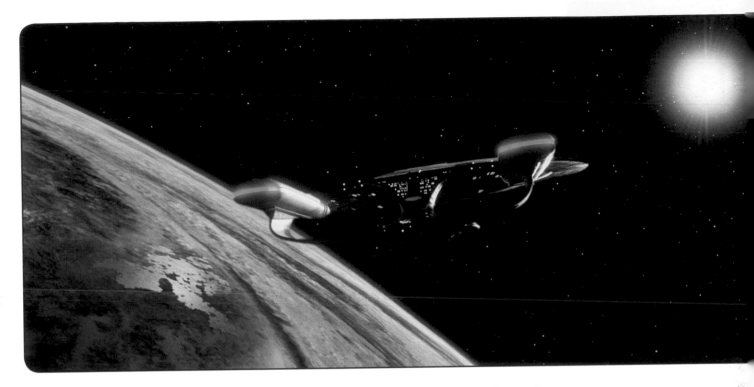

first Federation ships to successfully use a cloaking device. The technology had been developed in violation of the Treaty of Algeron 12 years earlier and used aboard the *U.S.S. Pegasus* NCC-53847 before the ship was lost in an asteroid field. The *Enterprise*-D was forced to use the interphasic cloaking device during its recovery after she became trapped inside a huge asteroid. The technology not only made the *Enterprise*-D invisible but also allowed her to travel through solid matter as her molecular phase inverter moved the ship out of phase with the space-time continuum.

For all the excitement and danger the *Enterprise*-D went through, most days were far more routine. In fact, according to Lt Commander Data, Stardate 44390 was an average day aboard the ship: this included, 'Four birthdays, two personnel transfers, two chess tournaments, a secondary school play, four promotions, the celebration of the Hindu Festival of Lights—and a birth and wedding.'

Unfortunately for the *Enterprise*-D, another day, this time in 2371, proved far from routine. While trying to save the Veridian system, the *Enterprise* was attacked by a Klingon Bird-of-Prey commanded by the Duras sisters. They modulated their weapons to the same frequency as the *Enterprise*'s shields, rendering them useless. Although the *Enterprise*-D destroyed the Klingon ship in the ensuing battle, she suffered heavy damage to her engineering hull, prompting an emergency saucer separation. The warp core breached moments later, blowing the stardrive section to pieces, and the resultant

shockwave hit the saucer section, knocking out its primary systems, including propulsion. Caught in the gravity of Veridian III the saucer section crash-landed on the planet's surface, and although there were no fatalities it was damaged beyond repair.

Surveying the wreckage later, Riker expressed his dismay over the fate of the ship that was once the pride of the fleet, but Picard voiced his opinion that he doubted it would be the last vessel to carry the name *Enterprise*. He was right. …

The crew of the *Enterprise*-D became heavily involved in Klingon politics and frustrated the Duras family's attempts to seize power.

of surrounding space or planetary bodies. The largest concentration of sensors was in the main sensor array, which was part of the main forward deflector.

These sensors included EM scanners, graviton scanners, life-form analysis scanners, subspace scanners, optical scanners, thermal scanners, and long-range scanners. Once the data had been gathered, it was processed and examined by an array of laboratories and departments throughout the ship. There were more than 100 research labs on board dedicated to a vast range of subjects, including stellar cartography, exobiology, astrophysics, cybernetics, archaeology, hydroponics, and geosciences.

If closer investigation of a planet was required, crewmembers could make use of the ship's six transporter rooms, eight emergency-evacuation transporters, and eight cargo transporters. They could also use one of the 37 shuttlecraft available, which were located in the ship's main shuttlebay in the saucer section, or in one of two smaller bays in the engineering

hull. The ship had a standard complement of 10 personnel shuttles, 10 cargo shuttles, 12 shuttlepods, and five 'special purpose' vehicles, plus additional shuttles as needed.

Small tractor-beam emitters were installed in the shuttlebays to help guide the shuttlecraft into the ship. The *Enterprise*-D was also equipped with a main tractor-beam emitter that was located on the under side of the rear part of the engineering hull. This projected a powerful graviton beam that could be used to hold an object in a fixed location or to tow another vessel.

All these departments, facilities, and systems required a massive amount of computing power to carry out their tasks. The computer system on board the *Enterprise*-D was isolinear based and enhanced by the Bynars, a race of computer specialists. There were three independent computer cores, two in the saucer section and one in the engineering hull. The data storage capacity for each core was 1.29024 x 10^10 kiloquads, running LCARS

The *Enterprise*-D was designed to spend extended periods in deep space so the crew were provided with extremely comfortable quarters. Senior staff had a living area equivalent to a small appartment on a planet's surface.

(library computer access and retrieval system) system software version 40274.

Of course, exploring the Galaxy can be dangerous, and it would have been foolhardy to send the *Enterprise*-D out armed only with diplomacy. In fact she was equipped with an impressive range of tactical systems. She had high-capacity deflector shields that could protect her from space debris and enemy fire, powered by ten generators throughout the ship. These deflectors could also be used to shield the ship from exposure to harmful radiation and other natural hazards. She also had 12 Type-X phaser banks, each capable of a 5.1 megawatt burst, and fore and aft torpedo launchers that could fire up to ten photon torpedoes simultaneously, each with an explosive yield of 18.5 isotons.

In the event of an injury to a crewmember, whether in a space battle or during the course of normal duties on board, they could be treated in the ship's impressive medical facilities that included the main sickbay on Deck 12. In addition to the main sickbay there were two other sickbay wards, 19 attached labs, and five operating theatres. Should there be mass casualties, the shuttlebays and the huge cargo bays on Decks 4, 5, 33, and 39 could be converted for use as triage wards.

The other main areas of the *Enterprise*-D were dedicated to crew quarters, and most were located in the saucer section, although there were also some in the stardrive section, which mainly housed the engineering crew. Crewmembers ranked lieutenant junior grade and higher had their own rooms, while those ranked below were normally required to share quarters. The senior officers had larger quarters that typically featured a living/work area and a separate bedroom and bathroom. The captain had the largest quarters, located on Deck 8, which was similar in size to the staterooms reserved for VIP guests.

Of course, not all members of the crew were human, or even humanoid. Decks 13 and 14, for example, were almost entirely given over to tursiops. Two years after her first mission, the *Enterprise*-D had 13 species aboard as part of her personnel, including Bajoran, Benzite, Betazoid, Bolian, El-Aurian, Klingon, Napean, and Vulcan.

In most parts of the *Enterprise*-D the environmental conditions were set for Class-M compatible, oxygen-breathing personnel, but the crew could also modify 10% of the living quarters to Class-H, K, or L conditions, with a further 2% capable of being modified to Class-N and N(2), even when the rest of the ship was Class-M.

The *Enterprise*-D was a truly enormous ship

that someone could live on for seven years and still not know their way around all of her parts and systems. When Chief Engineer Scott of the *U.S.S. Enterprise* NCC-1701 was briefly on board the *Enterprise*-D in 2369, he was dumbfounded by the advances and luxuries that had been introduced since his day. Even Captain Picard was 'in awe of its size and complexity', as the *Enterprise*-D represented the finest in starship design that the best minds in the Federation could produce in the late 24th century.

Tractor-beam emitters were located at the rear of the ship and were powerful enough to tow another vessel.

The *Enterprise*-D often provided emergency medical help to other ships, or even planets. Vast numbers of people could be beamed directly into the cargo bays.

The main bridge on the *U.S.S. Enterprise* NCC-1701-D was located on Deck 1 at the top of the saucer section, as it is on most Starfleet vessels. This egg-shaped room was the nerve-center of the vessel from where the commanding officers directed their missions and supervised the running of the ship's operations.

Even though the serious business of commanding Starfleet's flagship vessel was centered here, the bridge, like the rest of the ship, was decorated in calming, soft colors and furnishings to make it as welcoming as possible. It was much lighter and more spacious than the sparse, functional bridges on previous *Enterprises*, while it also did away with the constant sound of 'pings' and 'blips' that characterized earlier bridges, making for a much more relaxed atmosphere. Starfleet found that this 'calming' design increased crew efficiency and productivity.

The *Enterprise*-D featured a standard *Galaxy*-class bridge, but its modular design also made it much easier to replace if it became damaged beyond repair, or needed a system-wide upgrade.

The basic layout of the main bridge on the *Enterprise*-D retained many of the classic design elements of its predecessors, with the captain's chair at the center surrounded by the other command positions. This was because it made it easier for the captain to oversee all the key bridge personnel and issue them with orders.

Unlike earlier designs, the captain's chair was no longer isolated but was flanked by two other seats, with the first officer normally sitting to the captain's right and the counselor usually sitting to the left. This made consultation much easier, since by the 24th century command decisions had become a more collaborative process, even though the final orders still rested with the captain. The captain's chair featured miniaturized display screens and touch-sensitive control panels in the armrests, allowing direct access to some of the ship's most vital functions. The panel on the right arm of the chair allowed the captain to make log entries and gave access to intercoms, the library computer, and viewscreen control. The panel on the left arm featured controls for the piloting of the ship as well

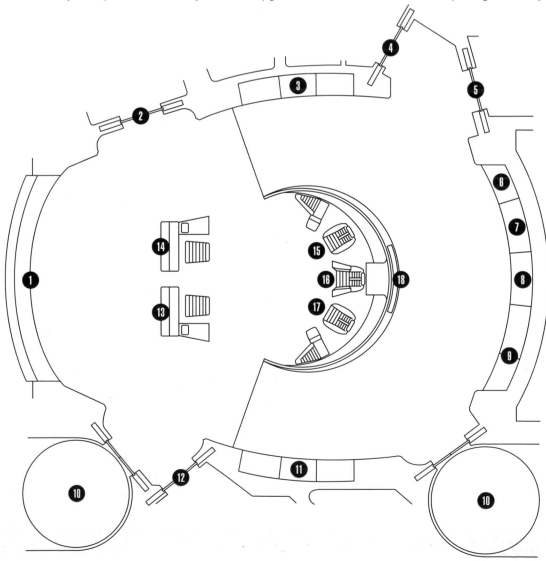

The main bridge was the nerve center of the ship. All of its functions were duplicated in a battle bridge in the stardrive section.

1. Main viewer
2. Turbolift to battle bridge
3. Isolinear subprocessors
4. Head
5. Exit to observation lounge
6. Science I
7. Science II
8. Mission operations
9. Environment and engineering
10. Turbolift
11. Isolinear subprocessors
12. Captain's ready room
13. Operations management
14. Conn
15. First officer's chair
16. Captain's chair
17. Counselor's chair
18. Tactical station

as its shields and weapons. Smaller backless chairs were located on either end of the command seats, where mission specialists or visiting VIPs could sit.

At the very front of the bridge was a large viewscreen that took up much of the front bulkhead wall and could be clearly seen by all bridge personnel. Its default mode was to display images in front of the ship relayed by the ship's sensors. However, it could provide views from anywhere around the ship, including above and below, while these images could be enlarged by a magnification of 10^6. Graphics and computer data re-routed from other stations could also be displayed or superimposed on an image to better illustrate information to the bridge crew. In addition, during communications the viewscreen could be used to display the real-time image of the person or life form the captain was talking to. The viewscreen could also display a 'red alert' status in case the loud alarm and flashing red lights were not enough.

Directly in front of the viewscreen were the operations (ops) and flight control (con) consoles.

Both these workstations featured chairs, which initially leant back in an almost prone position but were soon fixed in a more upright mode, while the touch-screen consoles swung aside to allow the crewmembers access to their seats. Facing the viewscreen, the ops station was on the port side and was often manned by Lt Commander Data. The main functions of this position were to oversee the running of the ship in terms of allocating resources, and to monitor communications and sensor systems. Ops was the management position of the *Enterprise-D* as it prioritized and coordinated the ship's resources, which was of particular importance if missions had conflicting or competing requirements. It was also the ops officer's responsibility to carry out the duties of a science officer, analyzing the data gathered by the sensors and acting on this insofar as it affected the running of the ship. All these duties required the ops officer to have a broad knowledge of science, technology, and engineering, not to mention starship operations and mission priorities.

OPS STATION

The Operations Management Officer, normally referred to as the Ops Officer, was responsible for coordinating a variety of the ship's functions, including the use of sensors.

MAIN BRIDGE

1. Emergency override
2. Departmental status display
3. Operational priority selection
4. Current action menu
5. Communications selection
6. X–Y transition pad control
7. Manual sequence controls
8. Conn control panel
9. Navigational reference display
10. Manual sequence controls
11. X–Y transition pad control
12. Warp-drive system controls
13. Impulse systems control
14. Emergency override

FLIGHT CONTROL STATION

The Flight Control Officer, normally referred to as the Conn, was responsible for piloting the ship and setting a course. He or she was in constant contact with engineering.

To the right of the ops workstation was the conn position, which combined the roles of the navigator and helmsman. It was from here that the *Enterprise*-D was piloted. The conn officer was responsible for plotting the course, setting the appropriate speed, and constantly monitoring the ship's course and heading. Inevitably, at high warp the navigational computers were responsible for providing course recommendations. The conn officer monitored the long-range sensors and adjusted course if they detected some obstruction. It was usually only in battle situations that the conn officer took full control of the actual piloting, though even then he or she still had the option of inputting a predetermined attack pattern.

The tactical console was located directly behind the captain's chair in a wooden rail and was manned from a standing position, usually by Lt Worf. This was the primary console for all the ship's defensive and offensive systems. As well as providing readouts on the internal security of the ship and warning of any intruders, it controlled the ship's defensive shields and the phaser and photon torpedo weapons. Other systems that were routinely controlled from here included communications, sensor equipment, and the ship's tractor beam.

There were several more workstations on the bridge, but these were normally manned only when additional crew were required to share an increased workload, such as in alert situations. When the *Enterprise*-D first entered service in 2364 the additional consoles ran along the rear wall of the bridge facing aft. As we have seen, from left to right these were science I, science II, environment, emergency manual override, and propulsion systems. By 2365 they had been reconfigured and mission ops had replaced emergency manual override, while engineering took the place of propulsion systems. Each of these workstations was normally operated from a standing position, but there were pull-out chairs below the consoles if needed. Sensibly, two food replicators were provided, allowing the crew access to a convenient source of sustenance without them having to leave the bridge.

The consoles around the rear of the bridge could be programmed to serve a number of functions. In the typical configuration they served as two science stations, mission ops, environment, and engineering stations.

A refit in 2371 saw the rear wall workstations reconfigured again—this time to science IV, mission ops, environment, and engineering I/II. Meanwhile, the storage and equipment lockers located on either side of the bridge were removed, and new consoles were fitted in their place. On the right side were science I, II, and III, while on the left were three communication stations. In addition, the command area—comprising the captain's, first officer's, and counselor's chairs—was raised to give a clearer view of the viewscreen over the ops and conn positions. New carpets and handrails were also fitted.

There were several access points to and from the main bridge. The main point of entry was to the rear on the right and next to this were two doors. One led to a bathroom (or 'head') while the other led to the observation lounge, where the senior staff often gathered to discuss the best course of action for their missions. On the left rear of the bridge was the aft turbolift and further round to the left was the door to the captain's office, or 'ready room', where the captain could work when not on duty and still be close to the bridge in case of emergency. To the left of the viewscreen was a door to a standard turbolift while to the right was another turbolift that led directly to the battle bridge so that senior crew could evacuate the main bridge and be at the ship's other main command center in seconds.

Befitting its status as one of the most important, if not the most important, areas of the ship, the main bridge had numerous environmental and power backups that allowed the bridge crew to continue working for 72 hours if there was a ship-wide power failure. So despite the fact that the main bridge was located in one of the most exposed parts of the ship at the top of the saucer section, it was, in fact, one of the safest places to be.

The tactical station was on a horseshoe-shaped rail behind the captain. It provided access to the ship's phasers and photon torpedoes and the ship's shields. It was also used to coordinate internal security.

The console next to the first officer's chair featured user programmable controls, provided data on the ship's status, and library computer access.

The engineering systems on board the *U.S.S. Enterprise* NCC-1701-D were dedicated to generating and controlling incredible amounts of power. The impulse engines alone could produce the kind of energy normally found at the heart of a sun; while the matter/antimatter reaction used in the warp engines generated even greater magnitudes of power. Most of this was used by the ship's propulsion systems—the impulse engines for sublight travel and the warp engines for faster-than-light travel.

Impulse power was the secondary power system on the *Enterprise*-D and was routinely used to propel the ship at a speed of 0.25c (one-quarter lightspeed), though it could achieve speeds of up to 0.92c (over nine-tenths lightspeed) when necessary. It was also used to power the ship's internal systems, including the computers and environmental systems. There were three impulse engines on the *Enterprise*-D, two in the saucer section located across Decks 9 and 10, and one in the stardrive section located over Decks 22 and 23.

Impulse engines work on much the same basic principle as old chemically fuelled space rockets—a quantity of mass is expelled from the ship, pushing the ship in the opposite direction in accordance with Newton's third law of motion. However, instead of being chemically fuelled, the *Enterprise*-D's impulse engines were powered by nuclear fusion. The main components in the *Enterprise*-D's impulse engines were fusion reactors that generated high-energy plasma, which was then passed through driver coils to create a subspace field that reduced the ship's mass, making it easier for the by-products vented through the engine's 'exhausts' to propel the ship.

If the impulse engines were needed to produce power for the internal systems rather than propulsion, the high-energy plasma was diverted from the driver coils to an electro-plasma system (EPS) distribution network that disbursed the energy as needed. Of course, it was possible to use the impulse engines to generate propulsion and internal power simultaneously.

The impulse engines, rather than the warp engines, were used to propel the *Enterprise*-D

1 **Master systems display console**

2 **Chief engineer's office**

3 **Duty engineer's station**

4 **Dilithium housing**

5 **Matter/antimatter reaction assembly**

ENGINEERING SYSTEMS

Main engineering on Deck 36 in the stardrive section was used to monitor the warp engines. It also provided direct control of all the ship's systems, which meant that it could act as a secondary bridge.

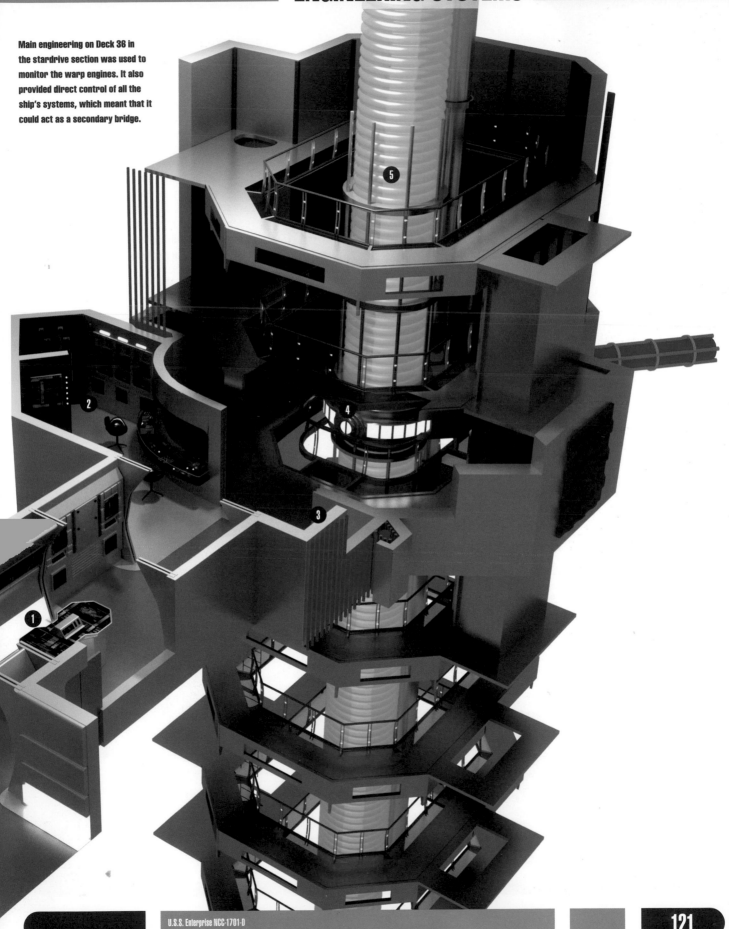

during maneuvering in relatively confined spaces, such as when docking at a space station or with another ship. Close maneuvering was accomplished by reaction control system (RCS) thrusters and mooring beams.

Impulse speeds were almost always used for travel within solar systems for safety reasons due to potential interaction with the gravity fields of planets and moons. Impulse speeds were also always used by the saucer section when the two parts of the ship separated, for the very good reason that the saucer did not have warp engines. Subspace coils on the saucer impulse engines could maintain a warp field for a brief time after separation from the stardrive section.

The *Enterprise*-D's primary power system was her warp engines, and the majority of her engineering systems were dedicated to generating and controlling this power. These systems were located over 12 decks in the stardrive section, with the most important being housed in main engineering, a multi-level facility located on Deck 36. This area was dominated by the warp core—or the matter/antimatter reaction chamber—a pulsing column measuring 36m high by 2.5m in diameter that glowed blue. It was this warp core that was at the heart of the ship's enormous power. Matter (deuterium) and antimatter (antideuterium) were injected into the warp core where they reacted, creating enormous amounts of pure energy.

The reaction of the matter and antimatter was controlled by the all-important dilithium crystals—the only known substance that does not react with antimatter. In order to achieve this non-reactive state it has to be energized with a high-energy EM field in the megawatt range. The plasma energy produced was then directed through a series of warp coils that were made from verterium cortenide and located in the ship's nacelles. Electromagnetic interactions between the plasma and the verterium cortenide coils then occurred to produce a change in the geometry of space surrounding the engine nacelles. In effect, this 'warped' the space surrounding the *Enterprise*, allowing it to travel at many times the

The master systems display console provided a constant stream of information about the ship's systems. It was tied into a wide range of internal sensors and provided feedback from independent self-diagnostic systems around the ship.

speed of light, while the ship itself was enveloped in a subspace bubble and never exceeded the speed of light. While each component of the warp engines was essential, perhaps the most important was the warp coils. Indeed, Commander Riker stated that warp coils were the most important invention of the past 200 years, as without them 'humans were confined to a single sector of the Galaxy'.

The maximum rated speed of the *Enterprise*-D was Warp 9.6—1,909 times the speed of light. At this speed it would take just 20 seconds to cross the solar system. This velocity was sustainable for approximately 12 hours before systems became critical, so was only used in emergencies. Warp 9.2 was the more normal maximum speed as it did not place such a strain on the engines, while Warp 6 was the average cruising speed.

Various facilities in main engineering were used to control and monitor the warp engines and their associated systems. The most important of these were the master situation monitor and the master systems display. The master situation monitor was a large computer display located on the forward bulkhead wall of main engineering that showed a large cutaway diagram of the ship and an overview of the status of the ship and its departments. If there was a problem with any one of the *Enterprise*-D's engineering systems, it would show up here. The master systems display was a large tabletop readout and control panel. Changes and recalibrations to the ship's engineering systems were often carried out here.

Dedicated displays to monitor the performance of the impulse engines were mounted on the wall near the master systems table, while on the opposite wall were the display monitors for the warp engines.

The chief engineer's workstation and several support consoles to his left and right were arranged around one side of the warp core behind a reinforced window. This allowed him to visually monitor the warp core and not just rely on the automated systems to alert him if there was something wrong.

The energy generated by the warp engines was also used to power any one of the *Enterprise*-D's 4,000 internal systems, including weapons, shields, computers, holodecks, and the environmental systems such as lighting, temperature, and gravity. The power was disbursed around the ship via electro-plasma system (EPS) conduits that had to be cleaned regularly to keep them working at optimum levels.

In the event of a serious malfunction, main engineering could be sealed off from the reactor core by isolation doors and containment force

Geordi La Forge became the *Enterprise*-D's chief engineer in 2365 after serving as the ship's helmsman. He stayed on the ship until she was destroyed by a warp core breach.

fields, but even these measures would not be enough if the warp core exploded. If that happened, it would blow the ship to pieces and destroy anything in the vicinity. If there was no way to stop the warp core from exploding, it could be ejected into space and hopefully the impulse engines would be able to take the ship far enough away in time to avoid the resultant blast. It would then have to be hoped that there was a nearby starship that could help, otherwise it could mean a very long journey back to the nearest starbase at impulse speeds.

The *Enterprise*-D was also outfitted with a large number of escape pods, which could be used if the ship suffered a catastrophic failure.

The consoles around the warp core provided access to all of the ship's engineering systems.

The *U.S.S. Enterprise* NCC-1701-D had 20 transporter rooms (standard configuration)—six main personnel transporters, eight cargo transporters, and six emergency transporters.

There were four personnel transporters on Deck 6 of the saucer section, two at each end of the vessel's core. Two additional personnel transporter rooms were on Deck 14 in the engineering section, again located near the center. All these rooms were near-identical, consisting of a transporter console, storage lockers located off to one side, and six individual transporter pads arranged in a hexagonal pattern with an oversized pad at the center for small cargo. Above the pads were molecular imaging scanners that scanned the person to be beamed on a quantum level while Heisenberg compensators took into account all the subatomic particles moving around inside the person. This amounted to billions of kiloquads of data. Phase transition coils then transformed the person into a stream of subatomic particles, called a matter stream, and stored it in a pattern buffer located below the transporter pad, where a biofilter screened out any harmful substances, including diseases. The matter stream was then sent via one of the 17 subspace emitter pads on the outer hull of the vessel to the required destination, where it was reassembled back into the person. During this process an annular confinement beam—a cylindrical force field—kept every part of the person within the transporter beam, ensuring none of their body parts were left behind.

The actual beaming process was quick, lasting just a few seconds. A person could not remain as a matter stream for longer than about 90 seconds as their molecular pattern would degrade and the signal would be lost. The signal had to remain above 50 per cent to be able to rematerialize the person. However, there was one astounding case when, thanks to some ingenious modifications to the transporter, Captain Montgomery Scott was rematerialized after being suspended in the pattern buffer for 75 years.

Enterprise-D's transporters were also capable of site-to-site transports, meaning a person could be transported directly from one location off the ship to another destination without having to materialize in the transporter room in between, though the transporter beam itself still returned to the ship's transporters before being redirected.

There were still limitations that the transporters of the *Enterprise*-D had not overcome. In general, they could not be used if the ship's shields were up or if there was a shield protecting the destination, though it was possible to beam through shields if the transporter chief had enough information about their frequency and modulation. It was also very strongly recommended that they not be used when the ship was at warp speed because of the severe spatial distortions caused by the warp field. Despite this, there was one instance of a successful transport being made while the *Enterprise*-D was at warp, but this was done as a last resort when the Borg had invaded Federation space, and even then both ships had to have matching warp velocities.

The cargo transporters were larger versions of the personnel transporters, typically featuring one large circular or oblong pad. There were four cargo transporters on Deck 4 of the saucer section and another four located in the cargo bay complex across Decks 38 and 39 of the stardrive section. All the cargo transporters were normally configured for low-resolution scans, meaning that they could only transport inanimate objects and not living beings. This also meant that they could handle more material in one go, because objects are far less complex than living beings and require far less memory and power to transport them.

In an emergency, the cargo transporters could be reconfigured to handle life forms and supplement the six emergency transporters used for evacuating personnel off the ship. There were four emergency transporters in the saucer section and two in the engineering hull, and each one could beam 22 people at a time. They could be used for beaming people off the ship, but not on to it, as they were fitted with scan-only phase transition coils. Their range was only 15,000km whereas the personnel and cargo transporters had a range of 40,000km.

All transporters required 87 seconds cool-down time before they could be used again. Using all 20 transporter rooms together, about 1,850 personnel could be beamed off the ship in an hour, meaning that the standard crew of 1,014 could be evacuated from the ship in less than 45 minutes.

Operation of the transporters appeared to be very straightforward and took just a few seconds, but both the hardware and software involved were extremely complex and even the slightest error in the transport process could result in death. To ensure the safety of their operation, the *Enterprise*-D had a dedicated transporter chief, who was responsible for the maintenance and repair of the transporters. For most of the *Enterprise*-D's operational life this was Chief Miles O'Brien, who often worked out of Transporter Room 3 and operated the controls himself.

As a result of stringent safety protocols and advancements in transporter technology, there were far fewer accidents involving transporters than there had been in the earliest days of transporter use. Nevertheless, there were a couple of unusual occurrences involving the transporters aboard the *Enterprise*-D. In 2368 Captain Picard, Ensign Ro Laren, Guinan, and Keiko O'Brien were transported back to the ship seconds before their shuttle was destroyed by a spatial anomaly. Portions of their pattern streams were obscured

during the transport and the ribo-viroxic-nucleic structures were removed from their DNA so that they rematerialized as 12-year-olds, but with their adult minds. Thanks to Dr Crusher the missing parts of their genetic codes were replaced using the transporter and they were returned to their former selves.

Another incident in 2368 led everyone to believe that Geordi La Forge and Ro had been killed in a transporter malfunction when beaming back from a Romulan ship. Their transporter patterns had, in fact, been corrupted by a Romulan cloaking device, sending them out of phase with normal matter and energy, in effect making them invisible and able to walk through solid objects. Geordi La Forge managed to work out what had happened to them and devised a way of having himself and Ro saturated with an anyon beam that dispersed the chroniton field around them, returning them to normal space and time. Lt Reginald Barclay even encountered quasi-energy microbes that were capable of living in a transporter beam.

Transporters were the standard method of traveling to and from the ship and were normally used to bring diplomatic visitors on board, in preference to docking with other vessels.

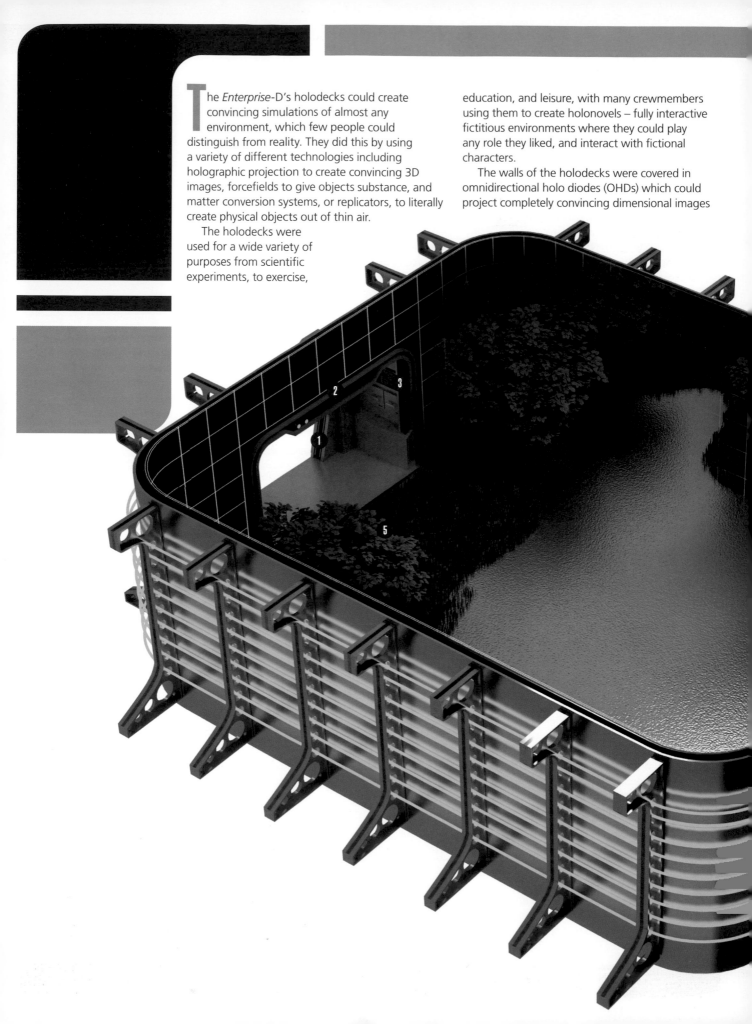

The *Enterprise*-D's holodecks could create convincing simulations of almost any environment, which few people could distinguish from reality. They did this by using a variety of different technologies including holographic projection to create convincing 3D images, forcefields to give objects substance, and matter conversion systems, or replicators, to literally create physical objects out of thin air.

The holodecks were used for a wide variety of purposes from scientific experiments, to exercise, education, and leisure, with many crewmembers using them to create holonovels – fully interactive fictitious environments where they could play any role they liked, and interact with fictional characters.

The walls of the holodecks were covered in omnidirectional holo diodes (OHDs) which could project completely convincing dimensional images and, if necessary, generate forcefields around them to create the sense that they were solid. The surface texture of the forcefields could even be manipulated so that, for example, the surface of a holographic tree felt like bark.

Typically, forcefields were only used to give foreground objects substance, while objects seen in the distance were actually holographic projections in two-dimensional space.

Any objects that were likely to be handled by the user were normally 'real'; that is, they were physical objects that were created using replicator technology. Thus if someone were to eat a meal in a holodeck, the food would be real and would satisfy their hunger. Under normal circumstances, computer safeties prevented the system from creating

physical objects such as bullets that could be harmful. Although objects that were replicated could theoretically be removed from the holodecks, the system was designed to make this impossible.

Sounds were created by hidden speakers, some smells were created by atomisers, others by replicators.

Holodecks could easily create the illusion that environments were much larger than the holodeck itself by moving the floor under the user's feet and manipulating the horizon to trick the user's sense of perspective.

The sophistication of a holodeck program was only limited by the amount of memory and computer processing power available. By the time the *Enterprise*-D entered service, more than enough power was available to create convincingly 'real' holographic characters, who were so sophisticated that they seemed to be sentient. In fact, on several occasions *Enterprise*'s holodeck systems appear to have created sentient life, in particular a holographic version of Professor Moriaty from Conan-Doyle's *Sherlock Holmes* novels. This character was not only self-aware but had emotions and free will.

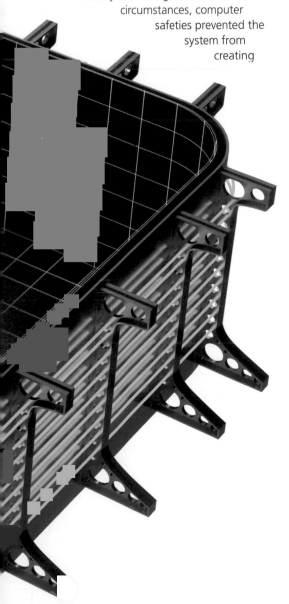

1 **Exit to corridor**

2 **Holographic arch**

3 **Virtual computer control panel**

4 **Omnidirectional holo diodes**

5 **Holographic projection**

The *Enterprise*-D's holodecks provided a complete and immersive virtual reality experience that was indistinguishable from the real thing.

The computer systems on board the *Enterprise* NCC-1701-D were incredibly powerful and did much more than just store vast amounts of information in the ship's library. They controlled just about every system on board, from food replicators to the warp engines. Computer control consoles were everywhere on the ship, so that a crewmember rarely had to walk more than a few strides to reach one; and even if they couldn't, they could access the ship's computer using vocal commands.

The performance of the computer systems on board the *Enterprise*-D was quite staggering. The memory capacity of the computers was 129,024 gigaquads, while the memory access speed was 4,600 kiloquads per second. Such were its abilities that the Bynars, a race of computer specialists, tried to hijack the ship so they could download the entire contents of their home world's computer into the *Enterprise*-D's system to stop the data from being lost when a solar flare threatened to destroy their planet.

The computer systems on the *Enterprise*-D were based on isolinear circuitry and isolinear optical chips, which had replaced duotronic computer technology around 2329. Isolinear chips were made of linear memory crystal material and employed holographic technology for both information storage and data processing. These chips were the basic building blocks of the computer's memory, holding both software and data routines; each chip could store 2.15 kiloquads of data.

Despite being ship-wide, the computer systems were concentrated in computer cores. There were three such independent computer cores on the ship: two in the saucer section and one in the engineering hull. All of the ship's essential computer processing functions could be handled by just one core.

The two cores located in the saucer section were vast cylindrical chambers that ran from Deck 5 to Deck 14 and housed the ship's primary computer hardware and processing equipment. The computer core in the engineering section was located between Deck 30 and Deck 37. The two cores in the saucer section normally worked in sync

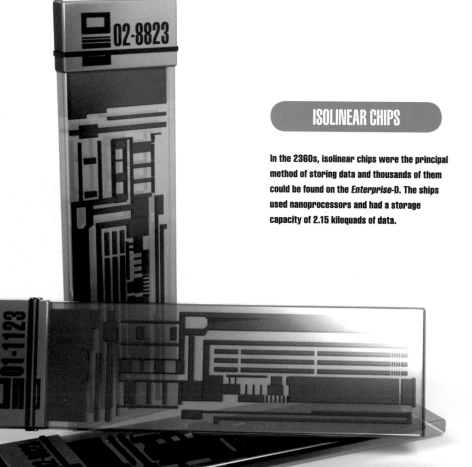

ISOLINEAR CHIPS

In the 2360s, isolinear chips were the principal method of storing data and thousands of them could be found on the *Enterprise*-D. The ships used nanoprocessors and had a storage capacity of 2.15 kiloquads of data.

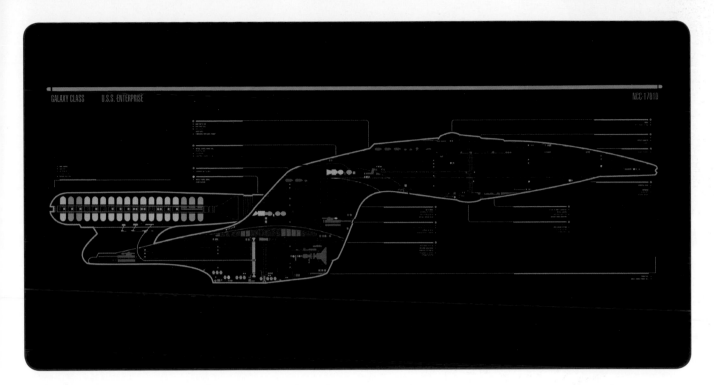

GALAXY CLASS   U.S.S. ENTERPRISE   NCC-1701D

with one another, and should one fail the other could immediately take on its load with little or no disruption to the ship's operations. The core in the engineering hull was a backup, used if a problem arose with both of the cores in the saucer section, or to run the stardrive section when the ship separated.

Each core incorporated a series of miniature subspace field generators that allowed data to be processed and transferred at faster-than-light speeds. Each one also had a systems monitoring room located at the top level where diagnostics and maintenance could be performed.

The computer cores interfaced with the ship's systems via an optical data network (ODN)—a large bundle of fiber optical cables that transferred data to the console panels all round the ship. All console display panels on the ship ran library computer access and retrieval system (LCARS) software. This meant that any console could be configured to operate any system on the ship, enabling, for example, the ops console on the bridge to run all the systems in engineering. Access was limited by a crewmember's security clearance rather than the console they were using.

All computer interfaces on the *Enterprise*-D were touch-sensitive console displays. They worked by having a sensor matrix embedded in them that detected pressure and gave tactile and auditory indications to confirm that controls had been activated.

The *Enterprise*-D could download and exchange information with another ship, or starbase, or

computer network on a planet without the need for physical contact via subspace transceiver arrays (STAs).

The computer systems were intrinsic to the smooth running of the *Enterprise*-D. While the crew were far from redundant, the computers automatically ran most of the ship's key systems, from regulating environmental controls to making the calculations needed for warp travel. Without them, the functioning of the ship would have been severely impaired and even unsafe, as was proved in 2366 when nanites—sentient microscopic robots—infected one of the computer cores, severely disrupting many of the ship's systems and putting it in danger. Fortunately, there were many fail-safes built into the computer systems. Indeed, there had not been a recorded case of complete systems failure since 2287.

The *Enterprise*-D's computer cores were absolutely essential. They not only controlled the warp and impulse engines but also the on board environment and atmosphere. Without them the ship was completely helpless.

At the top of each of the computer cores there was a systems monitor room where crewmembers could gain direct access to the data processors.

The *Enterprise*-D had an impressive array of medical facilities on board and one of the largest medical care facilities in the fleet, taking up much of an entire deck. There were three sickbays and four medical laboratories on board, but the primary sickbay was on the starboard side of Deck 12 in the saucer section.

This health facility comprised a surgical biobed in the centre of the room, with four more biobeds against the wall down one side that were used for patients under intensive care or in recovery. Separate, private rooms were also available for long-term patients.

The central biobed was used for most routine examinations and for complicated surgical operations, such as fitting an artificial heart. It was equipped with an overhead circular sensor cluster that could diagnostically scan the patient as well as project a force field to maintain a semi-sterile environment. Results of the bioscan could be displayed on the wall-mounted panel adjacent to the bed. This biobed could also be hooked up to a surgical support frame, or 'clamshell' as it

was more colloquially known. This device encased the patient and was equipped with bioscanners, life-support diagnostics, a cardiostimulator and other support functions for surgical procedures. It also created a fully sterile field to the extent that the patient did not even have to remove their clothes.

The secondary, recovery biobeds were equipped with unobtrusive biological sensors that continually monitored the patient and displayed their vital signs on a biofunction monitor above their head.

Other areas of main sickbay included the doctor's office and waiting area (it appears that even in the 24th century doctors couldn't keep to appointment times). The office was used by the chief medical officer to work in privacy or to conduct consultations with patients. This space could be personalized—for example, in *Enterprise*-D's sickbay a large painting in the waiting area depicted an abstract representation of humanoid organs against the background of space and several *Enterprises*.

1 Chief Medical Officer's office

2 Access to corridor Deck 12

3 General status display

4 Surgical support frame

5 Surgical biobed

6 Intensive care ward

7 Biofunction monitor

8 Recovery biobed

The *Enterprise*-D's sickbays provided everything that was needed by the thousand-strong crew, including GP's services and dentistry. It was also an advanced surgical facility that was the equal of any planetside hospital.

Other facilities surrounding sickbay included a biohazard ICU, isolation rooms for contagious patients, a morgue, and a physical therapy room for those undergoing rehabilitation. There was also a trauma stasis unit where critically injured patients could be placed in suspended animation until the ship arrived at a facility able to treat the injury.

The normal staffing level in main sickbay was four doctors, three medical technicians, and 12 nurses, although this was obviously subject to change depending on patient loads.

A smaller, secondary sickbay was located on the port side of Deck 12, while another one was found in the stardrive section. These contained the same equipment as the primary sickbay, but in different configurations.

The medical lab attached to main sickbay contained bio-assay and life-form analysis equipment as well as genetic sequence, nanotherapy, and virotherapeutic technology. These facilities were used mainly when studying a new life form or when creating a treatment for a new virus or unknown disease. Medical personnel could also run tests and monitor experiments here.

In the event of a medical emergency, with multiple casualties, auxiliary medical facilities could be set up in the shuttlebays, cargo bays, Ten-Forward, and in the guest quarters on Decks 5 and 6.

The *U.S.S. Enterprise* NCC-1701-D was designed primarily as a vessel for peaceful exploration, but she was also fully equipped with weapons and defensive systems. She was armed with 12 Type-X phaser arrays, three torpedo launchers, and a supply of at least 250 photon torpedoes, and was protected by a high-capacity deflector shield grid.

The *Enterprise*-D's primary weapons—and the ones used most often—were her phasers. These were located at various strategic points on the outer surface of the ship, and could cover 360°. The two main arrays nearly encircled the dorsal and ventral surfaces of the saucer section. The other arrays were all located on the stardrive section: two were on the upper rear surface of the interconnecting neck; two were on the upper surface of the warp engine pylons; two were on the lower surface of the warp engine pylons; two were on the port side and two on the starboard side on the lower surface of the warp engine pylons, near the upward bends of the pylon; one was on the ventral surface of the engineering hull; and the final one was hidden until the ship separated, when it

could be seen at the top of the interconnecting neck of the engineering module.

Each phaser array comprised a large number of phaser emitters in a strip that lit up before the energy phaser beam was directed towards its target. The dorsal phaser array on the saucer section consisted of 200 emitters, and with each emitter capable of a 5.1-megawatt burst it was an immensely powerful weapon. The effective tactical range of each phaser array was approximately 300,000km and all 12 arrays could be fired simultaneously in different directions. If need be, the ship's phasers could be reconfigured to drill into a planet's surface, which they did in 2370 when they bored right through to the core of Atrea IV in order to reheat it.

The only drawback of the phaser arrays was that they were normally only fully effective at sublight speeds. If the *Enterprise*-D needed to fire at warp speeds, the best option was to use photon torpedoes. The ship had three torpedo launchers: two in the engineering hull—one firing forward, the other rearward—and one launcher in the saucer

The *Enterprise*-D was a heavily armed and well-defended ship. Under normal circumstances she was protected by deflector shields but she was destroyed when the Duras sisters managed to obtain the frequency of these shields, allowing their weapons to penetrate them.

section. The main forward torpedo bay where the torpedoes were prepared for launch was located on Deck 25. Each launcher was capable of firing ten torpedoes at the same time, with each one being independently targeted.

The torpedo launchers aboard the *Enterprise*-D worked by using a combination of a launch-gas generator, an electromagnetic accelerator, and a sequential field of induction coils to propel the torpedo out of the ship in a warp field. The torpedo's own internal sustainer engines then maintained the warp field, allowing it to travel at faster-than-light speeds once clear of the ship.

The photon torpedoes themselves were black, lozenge-shaped casings that contained separate packets of matter and antimatter in their warheads. Only after launch were they mixed in a combiner tank, but still kept separate from each other in magnetic packets. Upon impact, the matter and antimatter met, causing a massive explosion. The photon torpedoes had an effective range of approximately 3,500,000km and the warhead yields had at least 16 preset levels. The maximum explosive yield was in the region of 18.5 isotons—an explosive yield of 25 isotons could destroy a large city. Each torpedo was fitted with a subspace detonator that could destroy it if it needed to be aborted after it had been launched.

The *Enterprise*-D's primary defensive system was her deflector shields, which used a type of force field to protect her from natural hazards and hostile fire. Without these shields the ship would be extremely vulnerable, as a single photon torpedo would be enough to completely annihilate it.

The deflector shield was created by ten generators located all over the surface of the outer hull, each producing 4.73 gigawatts, which created a localized spatial distortion that surrounded the ship. When the shields were up they used a variant frequency modulation to harmlessly deflect away most matter and energy. The frequency of the shields was a closely guarded secret, since if an enemy force were to obtain details of the frequency modulation they could modify their weapons to pass straight through them.

When the *Enterprise*-D's shields were up they were strong enough to protect her from the fire of most Alpha Quadrant starships, but continuous energy discharges could progressively weaken them. Each combat situation was different, making it difficult to know exactly how many hits would cause complete failure, but tactical officers continually reported the shield strength, expressed as a percentage. Often specific areas of the shields took more damage than others, so shield

power could be re-routed to provide additional reinforcement in these areas.

The other main systems involved in the *Enterprise*-D's weapons and defenses were the sensors and computers. The sensors could detect and track where and how far away the threats to the ship were, while the ship's computer could almost instantaneously calculate the most effective attack pattern or evasive maneuver to deal with the situation.

When the crew fought the Borg in 2366 they adapted the ship's navigational deflector to act as a giant phaser emitter.

Conventional phaser fire was of limited use against the Borg, who could adapt their defenses and repair damage with enormous speed.

The *Enterprise*-D had three shuttlebays and used a total of 72 auxiliary craft during her time in service. These comprised 12 Type-15 shuttlepods; four travel pods; 15 Type-6 shuttles; ten Type-7 shuttles; ten Type-9A cargo shuttles; five special-purpose shuttles; ten Work Bees; five Sphinx type M1 workpods; and one captain's yacht.

The main shuttlebay was huge and located across Decks 3 and 4 at the rear of the saucer section. The other two shuttlebays were located in the neck of the engineering hull, meaning that in separation mode both parts of the ship were equipped with shuttles. The main shuttlebay was considerably larger than the other two and was accessed by one roll-up door. It was so big, in fact, that if it was suddenly decompressed the resulting thrust would be sufficiently great to move the entire ship. However, an atmospheric containment field was normally in place around each set of shuttlebay doors, which allowed the shuttlecraft in and out while maintaining the Class-M atmosphere inside the bays. In other words, it was not necessary to depressurize and repressurize the bays every time a shuttle entered or left.

A dedicated shuttlebay officer and a flight deck officer oversaw shuttlebay operations, working out of a control booth that overlooked the main shuttlebay. They had to get clearance for all shuttle landings and take-offs from the ops officer on the bridge before they were initiated.

Although shuttles could be manually piloted in and out of the ship, it was normally an automated procedure carried out by a series of precision short-range tractor beams, mounted around the shuttlebay doors. These tractor beams could guide the shuttle out of the doors with far more precision that any pilot could. Shuttle landings were also performed by the same tractor beams, latching onto the shuttle at a distance of 350m

The Enterprise-*D carried a wide variety of shuttlecraft, which could be launched from any one of her three shuttlebays.*

and gently guiding it into the shuttlebay, ensuring that there was no risk of collision during maneuvers.

In 2368 Dr Crusher ordered all three of the *Enterprise*-D's shuttlebays to be converted into emergency medical triage centers in anticipation of the large number of casualties they expected after the *U.S.S. Denver* crashed.

Shuttlecraft were used for ship-to-ship journeys and for taking crew to planetary surfaces when interference prevented the safe use of transporters.

TYPE-15 SHUTTLEPODS

The Type-15 shuttlepods were the smallest shuttlecraft carried on the *Enterprise*-D, at just 3.6m in length and with a mass of 0.86 tonnes. In standard operating mode they had just two seats in the front and a small cargo area in the rear, but they could be configured to have just a single pilot seat at the front while the space at the back could be fitted with a bench to carry more passengers. They had three points of entry—two gullwing doors at the sides and a hatch at the rear.

The Type-15 shuttlepod had two 500-millicochrane impulse engines, also known as impulse nacelles, that were located on either side of the craft. It did not have warp capability, its engines being capable of propelling it to speeds of just 12,800 meters per second, making it suitable only for limited interplanetary travel. It also had eight DeFl 657 hot-gas RCS thrusters—also known as microfusion thrusters—that were for delicate maneuvers such as take-off and landing on a planetary surface. Deuterium was the primary power source, but backup power was supplied by three sarium krellide cells. It was only lightly armed with Type-4 phasers and a deflector shield, and had sensor arrays and independent navigation.

There were two variants of this model of shuttlepod: the Type-15A, which was slightly faster, and the Type-16, which had more powerful impulse engines at 750 millicochranes. Some of the names and numbers of the Type-15 shuttlepods on the *Enterprise*-D included *El-Baz* (05), *Ley, Onizuka* (07), *Pike* (12), and *Voltaire* (03).

Data stole the *El-Baz* while under Lore's control and piloted it through a transwarp conduit before landing it on a planetary surface. He also piloted the *Onizuka* to planet Tau Cygna V when it was necessary to evacuate a colony there.

Geordi La Forge was piloting the *Onizuka* back to the *Enterprise*-D after attending an artificial-intelligence seminar on Risa when he was abducted by Romulans and returned to the *Enterprise* under their control.

The *Pike* was destroyed while transporting Hytritium to the *Enterprise*-D. The explosion was initially thought to have been caused by pilot

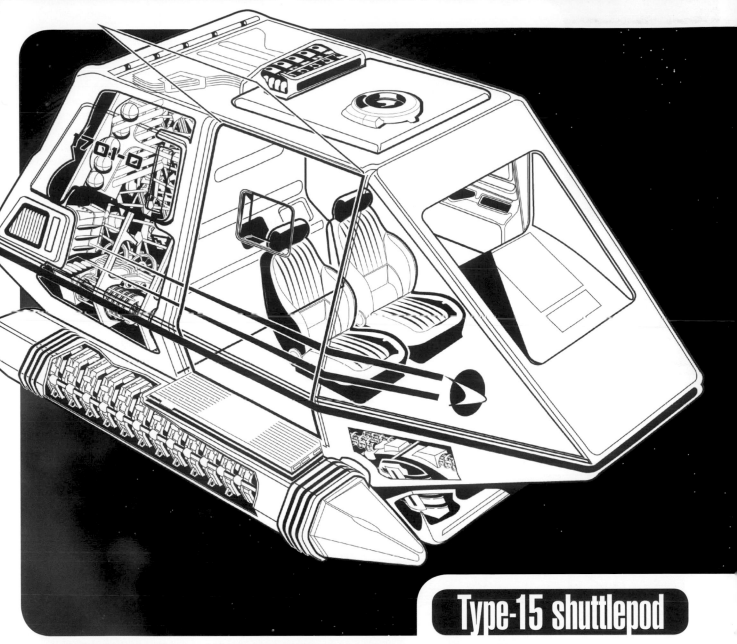

Type-15 shuttlepod

Classification	Type-15
Active service	2363—2371
Length	3.6m
Crew complement	2
Propulsion	Impulse engine
Weaponry	Type-4 phasers
Defenses	Deflector shield

error, but it was later discovered that Kivas Fajo was responsible.

Captain Picard piloted the *Voltaire* ahead of the *Enterprise*-D in order to lead it safely out of the Mar Oscura Nebula, which was riddled with rips in the fabric of space. The *Enterprise* made it, but the *Voltaire* was destroyed when it came into contact with one of these gaps. Picard beamed off just in time.

TYPE-6 SHUTTLECRAFT

Type-6 shuttlecraft were 6m long and could accommodate two crew at the front and up to six passengers on the bench seats in the rear. Unlike shuttlepods, the Type-6 did have warp capability and was equipped with a navigational deflector and two 1,250-millicochrane warp nacelles that allowed it to travel at Warp 1.2 for up to 48 hours. These factors made it much more suitable for longer missions than the shuttlepods. Type 6 Shuttlecraft were also fitted with 12 microfusion thrusters, RCS thruster quads, atmospheric airscoops, and hover field antigravs,

meaning they were just as well suited for flight inside a planetary atmosphere as they were in space.

Entry into Type-6 shuttlecraft was made via a large hatch door at the back that hinged at the bottom and provided a ramp when open, making it easier to load cargo on board. During their service aboard the *Enterprise*-D, Type-6 shuttlecraft were not normally armed, but they could be equipped with two Type-4 phaser emitters; they could also be fitted with a portable transporter system.

Type-6 shuttlecraft that served aboard the *Enterprise*-D included the *Berman*, *Curie* (03), *Fermi* (09/16), *Galileo* (07), *Goddard* (15), *Justman* (03), *Magellan* (15), and *Piller*.

TYPE-7 SHUTTLECRAFT

Type-7 shuttlecraft were similar to Type-6 shuttles, but they had more rounded hulls and were slightly longer at 8.5m. Like the Type-6, they were equipped with two 1,250-millicochrane warp nacelles, but on the Type-7 they were able to sustain a slightly faster speed of Warp 1.75 for 48 hours. The interior was also very similar to the Type-6,

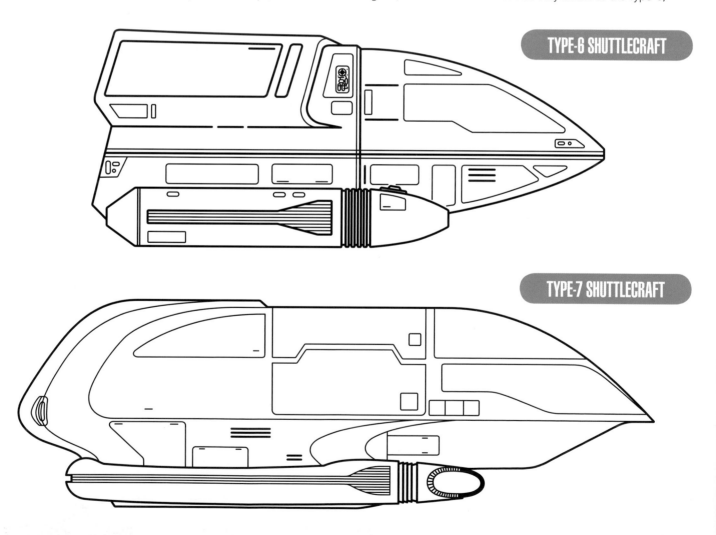

TYPE-6 SHUTTLECRAFT

TYPE-7 SHUTTLECRAFT

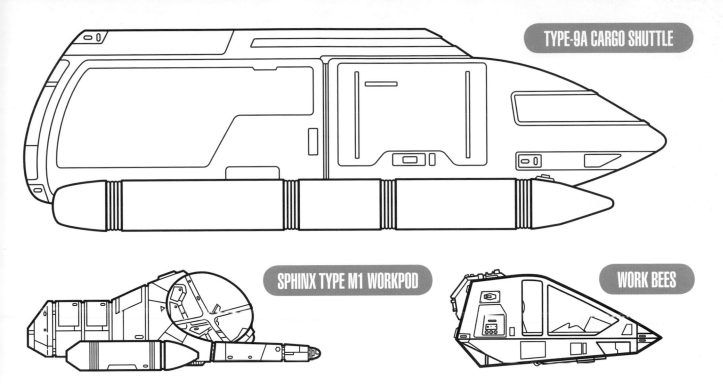

TYPE-9A CARGO SHUTTLE

SPHINX TYPE M1 WORKPOD

WORK BEES

except for a display panel between the two cockpit seats at eye level that gave navigational information and acted as a viewscreen. Again, like the Type-6 the Type-7 did not normally carry armaments, but could be fitted with Type-V phaser emitters.

The Type-7 had a boarding ramp in its nose. The pilot and co-pilot each had their own console that could be rotated forward to open the nose area for use of the ramp.

Type-7 shuttlecraft on the *Enterprise*-D included the *Copernicus* (03), *Feynman* (05), *Hawking*, and *Sakharov* (01).

TYPE-9A CARGO SHUTTLE

At 10.5m in length, the Type-9A cargo shuttle was the largest on board the *Enterprise*-D. As its name implies, it was mainly used for ferrying cargo and supplies and its shape was designed to maximize its load-carrying capacity. These shuttles were normally found working at Starfleet shipbuilding yards, where they could transport components from a planet surface to facilities in orbit. As they were warp capable, with two 1,500-millicochrane warp nacelles, they were often used on interplanetary cargo runs. On the *Enterprise*-D they were particularly useful for delivering medical supplies or other vital cargo to a planet while the *Enterprise* was engaged in her own separate missions.

The front of the Type-9A, where its two flight crew sat, was divided from the rear cargo area by a bulkhead wall, allowing the aft area to be opened in the vacuum of space if the need arose. This shuttle could also be converted to carry personnel and was particularly useful for transporting a large landing party and their equipment to planets that were heavily shielded, preventing the use of transporters.

SPHINX TYPE M1 WORKPOD

The Sphinx Workpod was similar to the Work Bee in that it was used to carry out repairs to the exterior of the *Enterprise*-D. However, at 6.2m the Sphinx was longer than the Work Bee, and it had a transparent, domed canopy enclosing the cockpit that allowed the pilot a much wider field of vision.

WORK BEES

Work Bees were small, yellow maintenance craft used to make repairs to the exterior of the *Enterprise*-D. They did not have warp capability and were powered by one microfusion reactor, giving them a maximum speed of 2,000 meters per second. They were piloted by a single person, who usually wore a spacesuit in case they had to exit the craft and perform repairs by hand.

They had large windows and a powerful light at the front to help the occupant see the damaged areas of the ship they were working on. The Work Bee could be modified with a number of tools, depending on the nature of the repairs it was carrying out, including remote grappling arms called a grappler sled, which folded beneath its body when not in use. It could also be fitted with a cargo-train attachment and be used for pulling multiple cargo containers.

Parallel universes

MULTIVERSE THEORY

One of the arguments that was regularly deployed to prove that time travel is impossible is that we haven't encountered anybody from the future. This led to the theory of parallel universes. Our normal concept of the universe is that every event has a single outcome. For example, if you toss a coin, it will come up either heads or tails, even though each outcome is equally likely.

According to the theory of the multiverse that's an overly simple way of looking at things. Instead it says that everything that *can* happen *does* happen. And that every time there is more than one possible outcome a new universe is created. So when you toss a coin, two parallel universes are created—one in which the coin comes up tails and one in which it comes up heads.

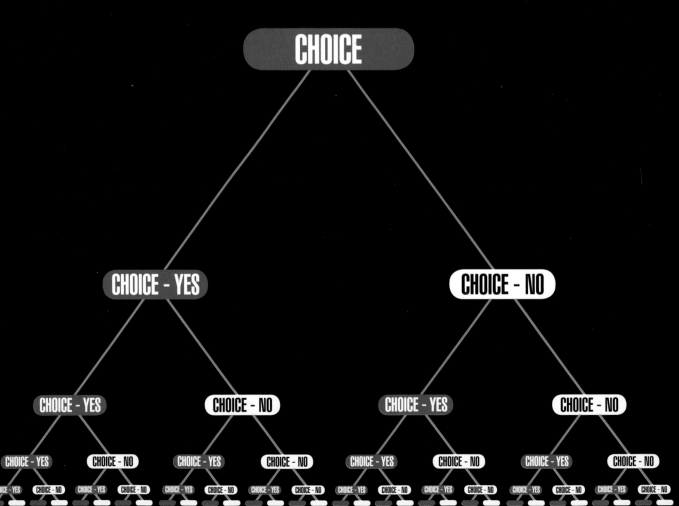

Theoretically this means that as soon as a time traveler arrives, they create a new universe, which isn't necessarily the same as the one that they left. So far Starfleet's experience of time travel is that it affects the travelers' own past, but the theory of the multiverse has been proved: in 2268 a transporter accident sent a landing party from the *U.S.S. Enterprise* NCC-1701 into a parallel universe that was similar to our own but had significant differences.

In this universe, the Federation did not exist. Instead, Starfleet was operated by the Terran Empire, which had conquered Vulcan and set out to conquer the rest of the Galaxy. Starships carried the registry *I.S.S.* rather than *U.S.S.*, but otherwise they were substantially the same and even had similar personnel. The 'mirror' *Enterprises* were laid out almost identically, though the 'Mirror Universe' version was more heavily armed and was fitted with torture chambers called agonizer booths, which were used to discipline the crew.

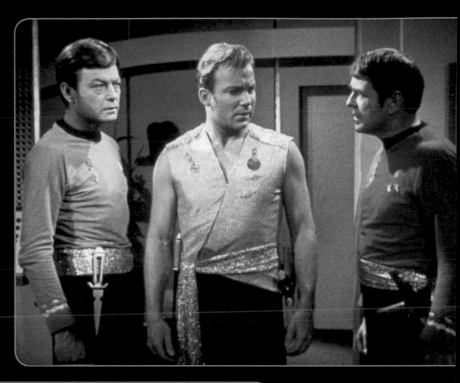

In 2268 a transporter accident sent Captain Kirk and several members of his senior staff into a parallel universe, in which the equivalent of the Federation was a brutal Empire focused on conquest rather than exploration.

In the 'mirror' universe, the *Enterprise* had an almost identical design and capacities. However, in this reality it was a warship. The crew were routinely armed even onboard ship, with officers murdering their superiors in order to gain promotion.

That original transporter accident involved an ion storm. The same parallel universe was also accessed over a century later, when a ship from *Deep Space Nine* entered the Bajoran Wormhole while suffering from a plasma leak. Somehow this caused the wormhole to act as a bridge between the universes.

The discovery of this 'mirror universe' proved the existence of at least one parallel

reality, but according to the multiverse theory there are actually an infinite number of universes or quantum realities. In fact, it suggests that every version of reality that one can imagine actually exists somewhere. There will be realities in which you are dead, realities where you are a king, and realities where you are a beggar.

Although it is impossible to prove the existence of every possible reality, this aspect of the multiverse theory was proved in 2370. On Stardate 47391.2 Lt Commander Worf of the *U.S.S. Enterprise* NCC-1701-D was returning from a *bat'leth* competition on Forcas III when he inadvertently piloted his shuttlecraft through a quantum fissure in the space-time continuum.

This fissure was a fixed point that existed in every quantum reality and as a result it was effectively a gateway between them. Worf's warp engines weakened the barriers between the different universes and he started to find himself shifting between parallel realities. At first he didn't realize what was happening since he was displacing his counterparts, but he soon realized that each universe was different to the last. In one universe he had won the *bat'leth* competition, in another he had placed ninth, in others he hadn't even attended. Eventually Worf found himself in universes where the differences were more pronounced. In several of the realities he visited, he was married to Counselor Troi, though in his original universe they were simply good friends.

With Commander Data's help, Worf realized what had happened. Everything that exists has a quantum signature that is unique to the quantum reality they belong to. Data discovered that Worf's quantum signature was different to everyone else's and that therefore he must be from a different universe. In this reality, Captain Picard had died fighting the Borg and Riker had become captain of the *Enterprise*-D, with Worf being promoted to become his first officer.

The crew devised a plan to restore him to his own universe by sending him back into the quantum fissure where he would restore the barriers between quantum realities by emitting an inverse warp field. However, the *Enterprise*-D was attacked while it was scanning the quantum fissure and this caused the barriers to break down even further. *Enterprises* from different realities started to appear in the universe Worf was trapped

in. Within minutes there were 285,000 *Enterprises* and the number was growing at a fantastic rate. The different crews were able to communicate with one another—in essence meeting themselves.

It now became urgent to restore the barriers between universes, which Worf did by completing the original plan. However, one of the *Enterprises* tried to prevent him doing so, because in their reality the Federation had been overrun by the Borg. To stop them, one Captain Riker was forced to fire on this version of the *Enterprise*-D, causing his own quantum double's death.

Fortunately, Worf was then able to seal the barriers between universes, restoring himself to his original quantum reality.

Theoretical studies by the Federation Department of Temporal Investigations have uncovered evidence that another parallel universe was created around 2233. This was apparently a cascade branching caused by events subsequent to the Romulan supernova of 2387. In this alternate reality, a rogue Romulan agent from the future caused the destruction of a Starfleet vessel, triggering a divergent timeline.

Notably, future *Enterprise* captain James T. Kirk, an infant at the time, was orphaned in the attack. This Kirk did nevertheless become a Starfleet officer, earning the position of *Enterprise* captain at an even younger age than his counterpart in the prime timeline.

The attack was the work of Nero, who blamed the Federation for failing to prevent the death of his wife and family in the explosion of the Romulan homeworld.

When Romulus was destroyed in the late 24th century, the Romulan Nero traveled back in time and massively altered the past. In the new timeline, the *U.S.S. Enterprise* NCC-1701 was a significantly larger ship than the one that Kirk commanded.

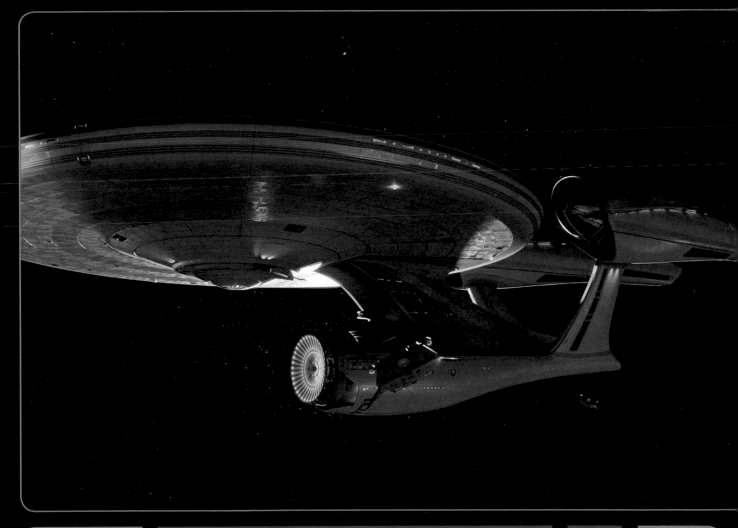

Classification	Sovereign class
Constructed	San Francisco Yards, Earth
Launch date	2372
Decommissioned	Still in service
Length	685m
Number of decks	24
Crew complement	1,500
Weaponry	Phasers and quantum torpedoes
Commanding Officers	Jean-Luc Picard

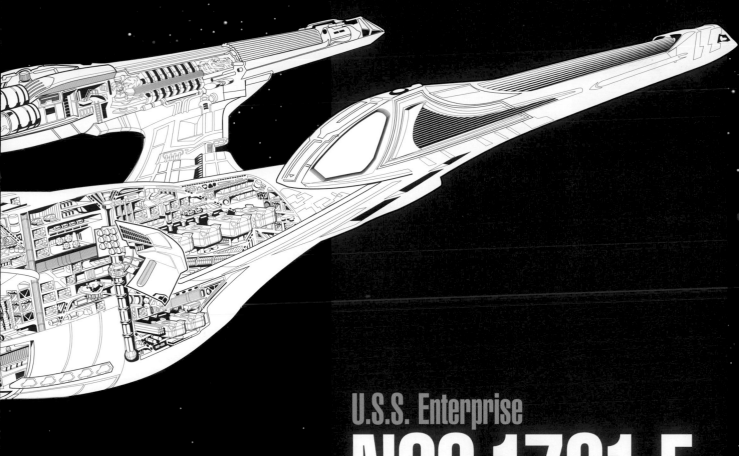

U.S.S. Enterprise
NCC-1701-E

The seventh Starship *Enterprise* was a *Sovereign*-class vessel that was launched from Earth's San Francisco Yards in 2372, under the command of Captain Jean-Luc Picard. The decision was taken to christen her *Enterprise* following the destruction of her predecessor in 2371. She played a major role in saving Earth from the Borg invasion of 2373, and six years later also saved the Federation from a major Romulan assault led by Praetor Shinzon. She ended 2379 undergoing a major refit that saw Captain Picard taking on a new generation of officers as many of her senior staff moved on to other positions.

SOVEREIGN CLASS

The *Enterprise*-E was a much leaner, tougher ship than its predecessor. Its design was heavily influenced by the Borg threat and as a result it didn't carry civilians. However, when the threat from the Borg passed it was given the same mission of exploration as all the previous *Enterprises*.

Following the loss of the *Galaxy*-class *Enterprise* NCC-1701-D in 2371, Starfleet opted to make the next *Enterprise* a *Sovereign*-class ship, and gave the name to a vessel that was already nearing completion. The senior staff of the *Enterprise*-D under Captain Jean-Luc Picard were assigned to the new ship, which was still under construction at the San Francisco Yards orbiting Earth. The *Enterprise*-E was launched in 2372 on Stardate 49027.5.

Like the *Enterprise*-D she was made Starfleet's flagship. Picard and his crew were among the most admired in the fleet. By this point Picard himself had turned down several opportunities to be promoted to admiral, feeling that he was best suited to commanding a ship that was on active service. His first officer, Commander Riker, had also passed up commands of his own in order to continue serving on the *Enterprise*. The rest of the senior staff consisted of Chief Medical Officer Dr Beverly Crusher, Chief Engineer Geordi La Forge, and Science Officer Data, all of whom had served on the *Enterprise*-D. They were joined by a new conn officer, Lt Hawk.

The only member of the senior staff who didn't transfer to the new *Enterprise* was the Klingon Commander Worf, who accepted a promotion and a new posting to *Deep Space Nine*, where his knowledge of Klingon culture proved invaluable. He later returned to service aboard the *Enterprise*-E after spending some time as a Federation ambassador to the Klingon Empire.

The *Enterprise*-E started her career with a one-year shakedown cruise in which all of her major systems were tested. In 2373 her mission was interrupted by the second Borg invasion. Initially Starfleet was concerned that Picard would be vulnerable because he had been assimilated by the Borg in 2366 (effectively becoming a member of this cybernetic species), so they ordered the *Enterprise* to patrol the Romulan Neutral Zone. However, once the battle began Picard and his crew disobeyed orders and joined the fleet that was engaging the Borg cube near Earth.

The *Enterprise*-E arrived at a pivotal moment— Admiral Hayes' ship had just been destroyed and the battle was in the balance. Picard assumed command

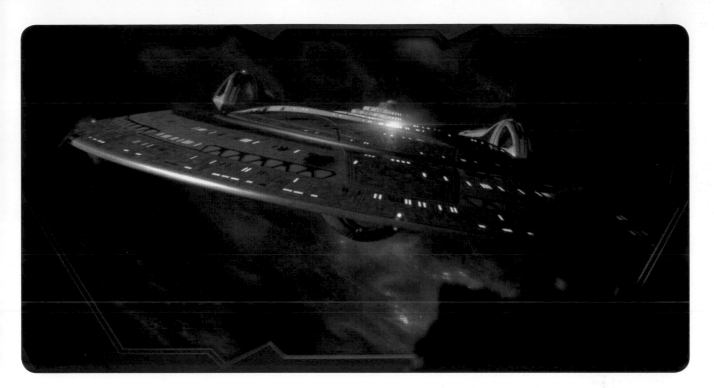

of the fleet and, using knowledge he had gained during his assimilation, managed to destroy the cube.

However, a group of Borg managed to escape and traveled into Earth's past, where they tried to prevent Zefram Cochrane's historic first warp flight and the ensuing first contact with the Vulcans. *Enterprise* pursued them and succeeded in protecting the timeline before returning to the present.

During her mission the *Enterprise*-E was partially assimilated by the Borg and had to spend several weeks in Spacedock before returning to active duty. She then served on the Cardassian front during the Dominion War and was involved in several distinguished missions.

In 2375, following the conclusion of the war, *Enterprise*-E was involved in exposing a plot to relocate the Ba'ku from their planet, where the atmosphere had incredible medicinal qualities.

In 2379 Starfleet sent the *Enterprise*-E to Romulus to negotiate with the new Romulan Praetor Shinzon, the leader of a Reman faction who had seized power. The *Enterprise* crew discovered that Shinzon was actually a clone of Captain Picard who had been created by an earlier Romulan regime as part of a plan to infiltrate the Federation. Changes in the Romulan senate had resulted in this plan being abandoned and Shinzon had been sent to the Reman mines, where he had grown up. As an adult he had organized a coup and made himself Praetor of the Romulan Empire. He claimed that he wanted peace with the Federation, but Picard soon realized that this was a ruse and that he planned to use a new weapon ship, the *Scimitar*, to destroy Earth.

The *Enterprise*-E engaged the *Scimitar* in the Bassen rift, and, with assistance from two Romulan Warbirds whose crews had rebelled against Shinzon, managed to defeat him. Picard only achieved this by ramming the *Scimitar*, causing significant damage to *Enterprise*. Commander Data was also killed during the mission when he disabled the Reman weapon.

The *Enterprise*-E returned to Earth Spacedock, where she was repaired and upgraded. The newly promoted Captain Riker took up his new posting on the *U.S.S. Titan*, where he was joined by Counselor Troi. He was replaced as Picard's first officer by Commander Martin Madden. Following her refit, the *Enterprise*-E was scheduled to begin a deep-space exploration mission to the Denab systems—with mission goals that were almost identical to her illustrious predecessor the *Enterprise* NX-01.

In 2379, the *Enterprise*-E prevented the Romulans from launching an all-out attack on Earth when she rammed headlong into the Romulan Praetor's ship, the *Scimitar*.

Opposite and below: The *Enterprise*-E was instrumental in defending Earth from the second Borg invasion. Captain Picard was able to use his unique understanding of the Borg to target the most vulnerable area of the Borg cube and destroy it.

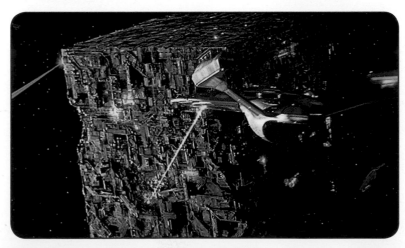

PORT ELEVATION

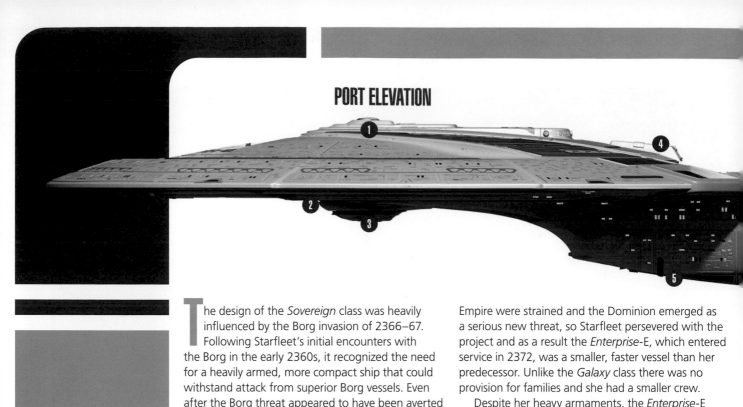

The design of the *Sovereign* class was heavily influenced by the Borg invasion of 2366–67. Following Starfleet's initial encounters with the Borg in the early 2360s, it recognized the need for a heavily armed, more compact ship that could withstand attack from superior Borg vessels. Even after the Borg threat appeared to have been averted in 2367, the Galaxy entered a particularly violent period during which relations with the Klingon Empire were strained and the Dominion emerged as a serious new threat, so Starfleet persevered with the project and as a result the *Enterprise*-E, which entered service in 2372, was a smaller, faster vessel than her predecessor. Unlike the *Galaxy* class there was no provision for families and she had a smaller crew.

Despite her heavy armaments, the *Enterprise*-E was still principally a research and exploration vessel, with a full complement of science labs and the

BOW ELEVATION

STERN ELEVATION

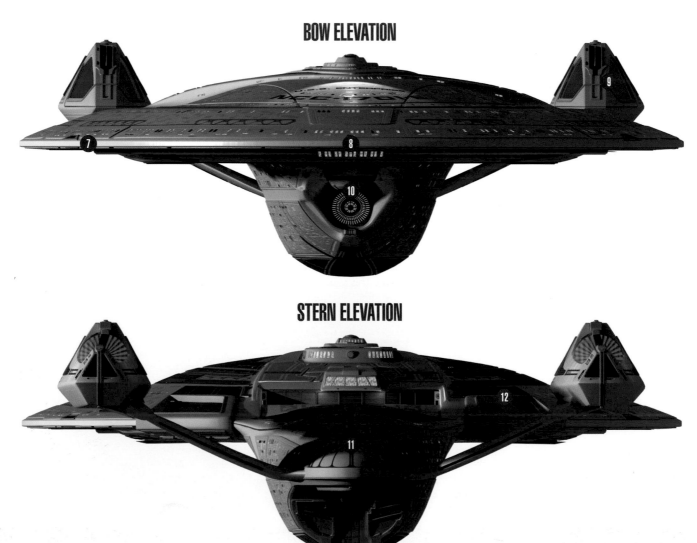

1. Main bridge
2. Torpedo launcher
3. Captain's yacht
4. Shuttlebay
5. Phaser array
6. Warp field grille
7. RCS thrusters
8. Saucer section navigational deflector
9. Bussard collector
10. Main navigational deflector
11. Shuttlebay
12. Saucer section impulse engine

appropriate mission specialists. However, because of the period in which she operated the *Enterprise*-E spent a significant amount of time involved in military operations. Consequently her weapons systems were upgraded several times, with Starfleet adding more phaser arrays and torpedo launchers. She used the latest weapons throughout her career, including quantum torpedoes and randomly modulated phasers, both of which had been developed to fight the Borg. During one of these refits the ship's interior was given a new configuration, increasing the number of decks to 29.

The *Enterprise*-E's more compact shape made her a more robust ship in combat, but although the structural elements were blended together it still followed the basic layout favored by Starfleet for over 200 years. The saucer had been elongated to become an oval but was still a distinct section

that contained the crew quarters and the majority of the science labs, while the secondary hull principally contained engineering systems. The twin warp nacelles were still supported by pylons and used super-heated plasma generated in a matter/antimatter reactor. The engineering and saucer sections were designed to separate and could operate independently of one another.

The *Enterprise*-E's engines were both more efficient and more powerful than their predecessor's. Following Dr Serova's realization in 2370 that warp technology was damaging the fabric of space, the ASDB (Advanced Starship Design Bureau) modified the Federation's warp engines to make them more efficient and to stop them causing permanent damage to subspace. The redesigns gave this *Enterprise*-E a cruising speed of Warp 8 and she could comfortably reach Warp 9.95.

Significant improvements were also made to the *Enterprise*-E's warp nacelles, most of which were designed to provide additional multiple redundancies and to increase safety. The new nacelles, which had 26 sets of warp coils, featured a fail-safe Bussard collector module. This part of the nacelle collected gasses from space and distilled them to harvest deuterium—the matter part of the matter/antimatter reaction that powered the warp engine. This new component allowed the collectors to function even when part of them was damaged.

The nacelles featured twin reaction control thruster assemblies for attitude control and subspeed maneuvers. The ASDB also added emergency plasma vents to the nacelle pylons. This allowed the engineering staff to vent plasma directly from the warp nacelles, removing the power that activated the warp coils. This feature reduced the likelihood of a warp core overload that would force the crew to eject the core.

In 2375 Commander Riker made innovative use of these design features to destroy a Son'a vessel that was pursuing the *Enterprise*-E. He used the Bussard collectors to scoop up metreon gas, which he then vented from the nacelles. To the Son'a it looked as if the *Enterprise* had been badly damaged and was venting warp plasma, so they fired a torpedo in an attempt to destroy her. As Riker had anticipated, the torpedo detonated the gas, destroying one of the pursuing vessels and crippling the other.

The *Enterprise*-E relied on her transporters for most missions but also carried a wide range of auxiliary vehicles, from standard shuttles to longer-range scout vessels, and all-terrain vehicles used to explore a planet's surface. In 2379 she took delivery of a new design of multipurpose shuttle, the *Argo*. In keeping with other starships of the period, the *Enterprise*-E had two shuttlebays—one at the rear of the saucer and the other at the rear of the engineering hull with traditional clamshell doors.

COMPUTER SYSTEMS

The *Sovereign*-class computers used the latest bio-neural circuitry, which was introduced in the

DORSAL ELEVATION

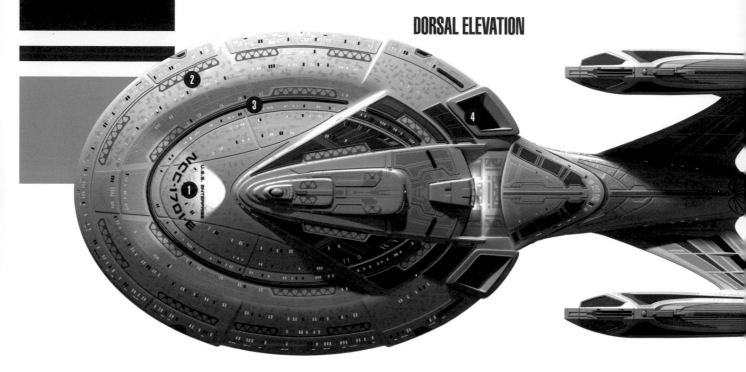

VENTRAL ELEVATION

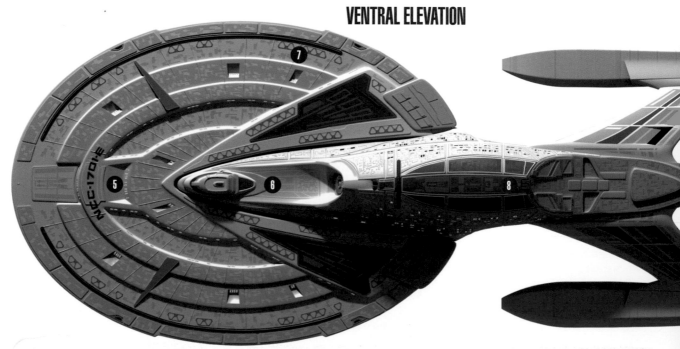

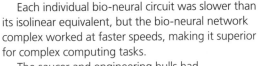

late 2360s. This technology used synthetic cells to process data far faster than traditional isolinear optical circuitry. The new systems could process 6,200 kiloquads of data a second. This semi-organic computer system still ran alongside a conventional ODN (optical data network), in part because bio-neural technology was so new but also because it was vulnerable to biological attack. Working together the two systems produced significant improvements in processing power and speed.

Each individual bio-neural circuit was slower than its isolinear equivalent, but the bio-neural network complex worked at faster speeds, making it superior for complex computing tasks.

The saucer and engineering hulls had independent computer cores that were linked by the ODN but could operate completely independently when the ship was in separated mode. The saucer section computer core ran between Decks 6 and 9; the engineering section computer core ran between Decks 17 and 19.

CREW FACILITIES

The crew quarters were all located in the saucer section, with the majority of the officers billeted on Deck 2. Although the ship was designed with combat in mind, it was equipped with all the luxuries a Starfleet officer could expect in the late 24th century. Food was provided by replicators, with individual units in each crewmember's quarters. The crew could also meet in a lounge, which, as a tribute to the bar on the *Enterprise*-D, was known as Ten-Forward. The senior staff also had a separate mess hall on Deck 2.

The ship was fitted with holodecks that could create a completely convincing replica of any place by using a combination of holograms, replicator technology, and moving floors. Although they had scientific applications and were used for education, training, simulations, and physical fitness, the holodecks were also used for leisure, with the crew adopting roles in holonovels that allowed them to act out their fantasies.

MEDICAL FACILITIES

The *Enterprise*-E's main sickbay had a standard design that could be found on several different starships of the period, including the *Intrepid* class. It was arranged into four distinct areas: a recovery ward with three standard biobeds; a circular diagnostic and surgical area, with a full surgical biobed that could generate a sterile field for operations, and high-resolution medical scanners for in-depth analysis and autopsies; a circular doctor's office; and a small lab.

The *Enterprise*-E was the first *Enterprise* to be fitted with an EMH (emergency medical hologram). This was a holographic doctor that had been programmed with the combined knowledge of 47 doctors and the full Starfleet medical database. In order for the EMH to function, sickbay was fitted with holographic emitters and force-field emitters that could give him substance so that he could touch his patients. The EMH was only intended for use in emergencies, and was not regarded as a replacement for traditional doctors.

1 Ship's registry

2 Escape pod hatch

3 Phaser array

4 Saucer section impulse engine

5 Saucer section navigational deflector

6 Captain's yacht

7 Escape pod hatch

8 Warp core ejection hatch

The main bridge occupied its traditional position on Deck 1, at the top of the saucer section, with the deck numbers then increasing as they went down the ship. The layout of the bridge followed the basic design of the *Enterprise*-D. The captain's chair was in the center of the room with seating to his left and right for his first officer and any mission specialist he needed to consult. The latter seat was normally used by the ship's counselor. In a departure from the previous design, the horseshoe console behind the captain's chair had been broken up. The tactical station was now to the starboard of the captain while the equivalent console to the port was normally assigned as a science station.

The conn and ops stations were still next to one another, directly in front of the captain. As on the original *Enterprise* NX-01, the helm officer could opt to pilot the vessel using a joystick rather than standard computer interfaces. In another design departure, which can be seen on other bridges of the period, additional consoles around the perimeter of the bridge faced into the room.

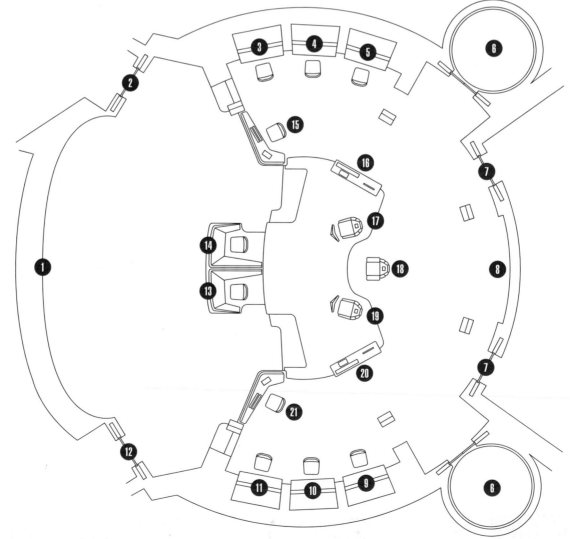

One obvious innovation was that the main viewer was projected holographically and could be switched on and off, whereas on previous *Enterprises* it had defaulted to showing the space immediately in front of the ship. When the viewer was off a blank stretch of bulkhead could be seen behind it.

The rest of Deck 1 consisted of a captain's ready room and an observation lounge.

An auxiliary bridge (also known as the battle bridge) was located on Deck 14 in the engineering hull. If the main bridge suffered catastrophic damage or was compromised, control of the ship could be re-routed to here. It also served as the bridge for the engineering hull in separated mode.

A service access port in the floor provided emergency egress in the event of turbolift failure. Also, on either side of the bridge was a replicator terminal, to provide refreshments to on-duty crewmembers.

1 **Main viewscreen**

2 **Captain's ready room**

3 **Science station I**

4 **Environmental controls**

5 **Engineering station I**

6 **Turbolift**

7 **Conference room**

8 **Master systems display**

9 **Science station II**

10 **Mission ops**

11 **Engineering station II**

12 **EVA airlock**

13 **Operations management**

14 **Conn**

15 **Science station**

16 **Tactical I**

17 **First officer's chair**

18 **Captain's chair**

19 **Counselor's chair**

20 **Tactical II**

21 **Engineering station**

When she entered service, the *Enterprise*-E had substantially the same senior staff as her predecessor, under the command of Captain Picard.

The warp core—a cylindrical vessel in which matter and antimatter react in a controlled manner to generate the high-energy plasma used to energize the warp coils—was almost exactly in the mid-point of the ship and ran the full height of the engineering hull from Decks 10 to 24. Main engineering was larger than on previous *Enterprises* and covered three decks, with the matter/antimatter reaction chamber and the plasma transfer conduits in the middle of these on Deck 15, and the main control room on the lowest of the three decks on Deck 16.

The majority of the control systems, including a master systems display console, were on the ground level with a gantry running around the warp core on Deck 15 to allow the crew to inspect it and to make adjustments. The warp core was flanked by plasma coolant tanks, which were used to control the temperature of the plasma that was transferred to the nacelles. This coolant was extremely dangerous since it would destroy any organic substances on contact.

The antimatter storage pods were located at the bottom of the ship on Deck 23, with

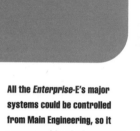

All the *Enterprise*-E's major systems could be controlled from Main Engineering, so it was natural for the Borg to take control of it when they invaded the ship. They rapidly converted it into a Borg command center, where they attempted to persuade Picard to join them.

hard-connect points in the ship's hull allowing them to be supplied directly from a refueling station. The deuterium tanks were on Deck 14, just aft of main engineering.

The warp core was designed to be ejected in an emergency. This procedure was actually performed in 2375, when the Son'a attacked using an isolytic subspace weapon that formed a tear in space which was attracted to the *Enterprise*-E's warp core. Commander La Forge collapsed the tear by ejecting the ship's warp core and detonating it.

WARP CORE BREACH

An uncontrolled reaction in the warp core poses a major threat to any starship—the explosion has the power to completely vaporize the ship—so in extreme emergencies Starfleet vessels are designed to eject the warp core. This separates the warp core from the matter and antimatter streams, removing the sources of the warp reaction, often meaning that the warp core can be retrieved at a later date.

On the *Enterprise*-E the warp core could be ejected through a hatch on the underside of the starship. The order to eject could only be given by a senior officer, normally the Chief Engineer. (In the event of massive casualties, the computer system would keep track of seniority among surviving personnel.) In certain circumstances the computer would automatically eject the warp core to protect the crew.

When the ejection system is activated, where possible fuel and power transfer conduits are sealed and separated from the warp core. The computer then blows out the warp core hatch and explosive charges force the warp core out.

If the core is still functioning normally, the crew can still control the matter/antimatter reaction, causing it to detonate remotely. This happened in 2375 when Commander La Forge ejected the warp core and then remote-detonated it in order to seal a subspace tear created by an illegal Son'a isolytic weapon.

The *Enterprise*-E could also eject the antimatter containment pods if they were in danger of breaching.

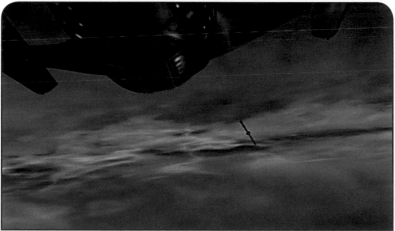

In 2375, the *Enterprise*-E was forced to eject its warp core to seal a subspace breach created by an isolytic weapon. The tactic was successful but the ship was unable to travel at warp until a new core had been installed.

The *Sovereign* class was always intended to be a fighting ship as well as an explorer and was heavily armed. When she was launched the *Enterprise*-E carried 12 Type-XII phaser strips with an output of 7.2 megawatts per emitter element. The primary phaser arrays were located on Deck 4, on the upper surface of the saucer, with another set of arrays on the under side; internally the phaser bank generators were on Deck 5. The Type-XII phaser had been modified so the beam modulation could be changed in either a predetermined or random sequence to make it more effective against Borg shields.

The *Enterprise*-E's own shields had a higher energy capacity than their predecessors and had also been designed to be re-modulated regularly to make them more resistant to Borg weaponry.

The most significant weapons upgrade involved the torpedoes. In addition to her complement of photon torpedoes the *Enterprise*-E carried the latest quantum torpedoes, which could be fired at warp speed and had a far greater yield than their predecessors. They accomplished this by using zero-point energy to produce an explosion that worked on the same principles as the original big bang that created the universe. The yield of a quantum torpedo is more than double that of a conventional photon torpedo.

Enterprise-E was initially fitted with five torpedo tubes arranged in two torpedo launchers, the forward one positioned on the under side of the saucer section just above the captain's yacht, with the aft one at the rear of Deck 24. Each of the launchers could fire a spread of 12 torpedoes. Tractor beam emitters were located alongside each of the launchers.

In 2376 the *Enterprise*-E underwent a major refit that increased the number of phaser arrays to 16 and added another five torpedo tubes. The additional phaser arrays were located on the warp nacelle support pylons, which were also replaced during the refit. Four of the new torpedo tubes faced aft, including a twin launcher behind the bridge and a single launcher above the hangar bay; an additional tube was also fitted to the front of the saucer.

The *Enterprise*-E was designed to fight the Borg and was given more advanced weaponry than her predecessor. This proved invaluable when she engaged and destroyed a Borg Cube that was attacking Earth.

QUANTUM TORPEDOES

Quantum torpedoes are able to produce an explosive yield of over 50 isotons, which is more than double that of a conventional photon torpedo. They do this by extracting energy from an area of space-time known as the zero-point vacuum.

Experimental work with zero-point energy began in 2236. It works by manipulating space-time to literally create new subatomic particles, a process that releases vast amounts of energy. In fact, a zero-point energy explosion is basically a tiny version of the big bang that created the universe.

The process requires a zero-point field reaction chamber. On a quantum torpedo this is 0.76 meters long with a diameter of 0.35 meters. The energy field inside is manipulated by jacketing layers of dilithium and synthetic neutronium. The zero-point field is initiated by a combination of an EM rectifier, a waveguide bundle, a subspace field amplifier and a continuum distortion emitter, which together make up the zero-point initiator.

The initiator is powered by a conventional matter/antimatter explosion, although this is slightly larger than that found in a conventional photon torpedo. When this is detonated, the zero-point initiator channels it into the reaction chamber creating new subatomic particles and causing the explosion.

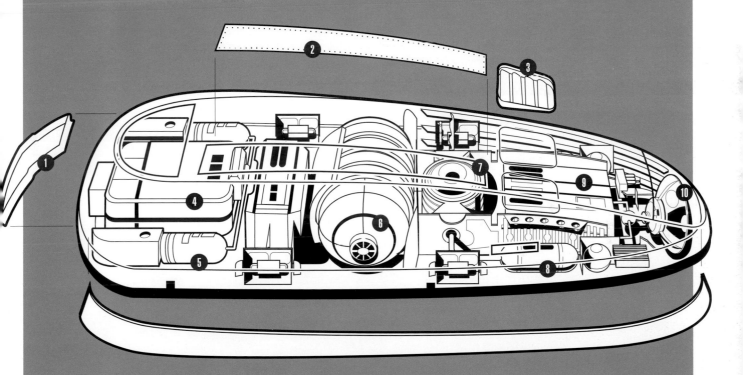

1. **Baffled propulsion vent**
2. **Service access**
3. **Thermal control vent**
4. **Sustainer propulsion module**
5. **Propulsion and thrust vector subsystems**
6. **Zero-point field reaction chamber**
7. **Positron accelerator**
8. **Wide-field proximity sensor**
9. **Guidance and targeting processor**
10. **Primary targeting scanner**

Like its predecessor, the *Enterprise*-E also had an auxiliary vessel, known as the captain's yacht, which was docked to the underside of the saucer section. At over 33 meters long, this vessel, the *Cousteau*, was more spacious than a conventional shuttle and was designed for diplomatic missions. It was large enough to host diplomatic parties and could be used to provide a neutral venue for negotiations.

It was capable of warp-speed travel and a small warp core ran along the ship's spine using the same architecture as the *Danube*-class runabout. Under normal circumstances the warp core provided power to all of the *Cousteau*'s systems, but it was also fitted with a small fusion engine, which powered the impulse engines and provided an auxiliary power supply in case of emergencies. Like the *Enterprise*-E's standard shuttlecraft, the *Cousteau* was able to operate in an atmosphere and could land on the surface of a planet. It could be piloted by a single crewmember, who could control all the systems from the cockpit at the front of the vessel. It was only lightly armed and had minimal shielding but was highly maneuverable.

When it was docked with the *Enterprise*-E it was indistinguishable from the main part of the ship and the crew could even walk straight on to it from a corridor on board the *Enterprise*. The *Cousteau* was also fitted with a transporter, which could be found in the ship's cargo area. In emergencies, it could be used to evacuate any personnel to a nearby location.

When the docking clamps were released, the *Cousteau* used reaction control thrusters to move clear of the main ship before deploying its warp nacelles, which swung down into place.

While the yacht was away from the *Enterprise*, the structural integrity and inertial dampening fields were adjusted to compensate for the new shape and the ship's performance was unaffected.

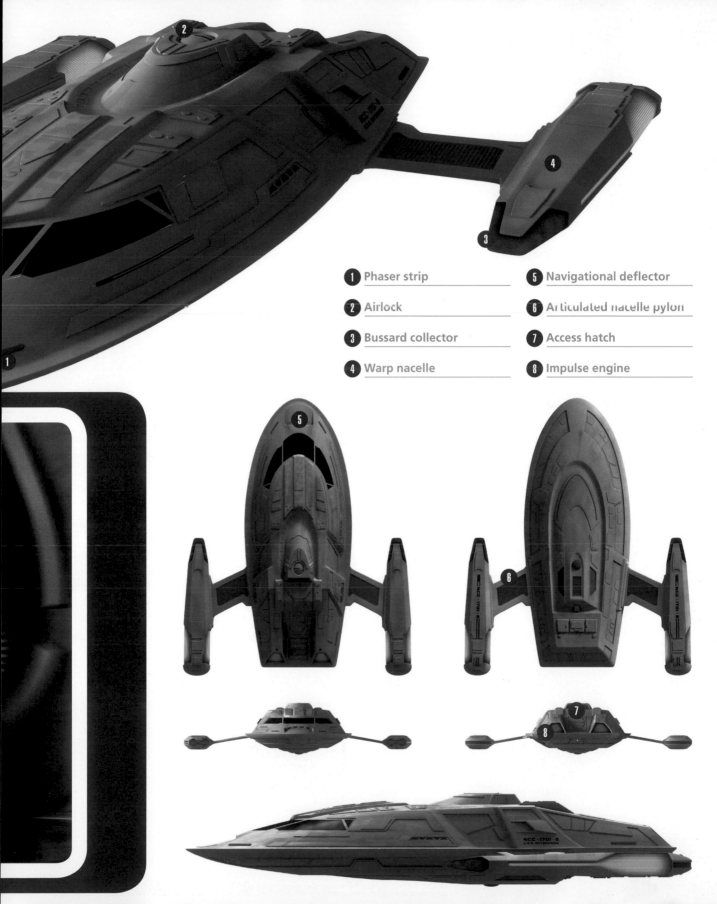

1 **Phaser strip**

2 **Airlock**

3 **Bussard collector**

4 **Warp nacelle**

5 **Navigational deflector**

6 **Articulated nacelle pylon**

7 **Access hatch**

8 **Impulse engine**

Appendix
SIZE COMPARISON CHART

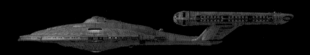

NX-01 Length: 225m

NCC-1701 Length: 298m

NCC-1701-A Length: 305m

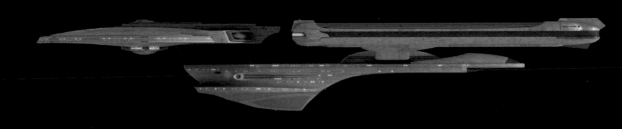

NCC-1701-B Length: 467m

NCC-1701-C Length: 526m

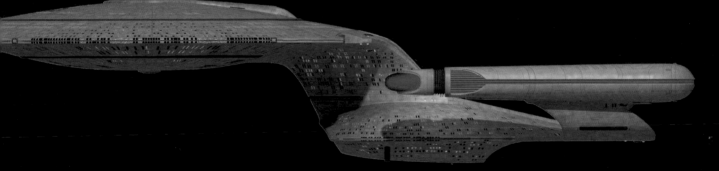

NCC-1701-D Length: 641m

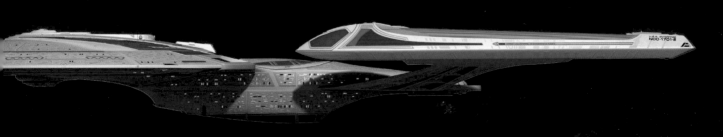

NCC-1701-E Length: 685m

ACKNOWLEDGEMENTS

The authors would like to thank Hilary Newstead for her sympathy and support, Robert Bonchune for providing the closest thing you can get to real starships, Doug Drexler for going the extra mile to tell us about the hidden aspects of the NX-01, John Eaves for sharing the thinking behind the *Enterprise*-E, John Van Citters, Marian Cordry, and Risa Kessler for their excellence in all things involving licensed publishing, and various members of the *Star Trek* writing staff for the interviews they have given over the years—it seems wrong to single people out, but particular thanks go to Ira Steven Behr, Hans Beimler, Ronald D. Moore, Joe Menosky, Dorothy Fontana, Andre Bormanis, and Bryan Fuller.

Specific thanks must go to the following people, who have played important roles in the design of the *Enterprise* over the years:

Gene Roddenberry, Michael Piller, and Walter M. "Matt" Jefferies, who started it all with his brilliant design for both the interior and the exterior of the original *Enterprise*. His genius helped shape Gene Roddenberry's vision of tomorrow.

Andrew Probert, Richard Taylor and Douglas Trumbull designed the refit Enterprise, based on Matt Jefferies' Phase II *Enterprise*. Production designers Harold Michelson and Joseph R. Jennings were responsible for that ship's interiors.

Andrew Probert also designed the *Enterprise*-D exterior and that ship's bridge for production designer Herman Zimmerman. Fellow production designer Richard D. James was responsible for the *Enterprise*-D for the second through the seventh seasons of *Star Trek: The Next Generation*.

Rick Sternbach designed the *Enterprise*-C model for the fan-favorite episode *Yesterday's Enterprise* while Richard D. James designed that ship's interiors.

Herman Zimmerman and Jack Collis were responsible for the *Enterprise*-A interiors in *Star Trek IV—Star Trek VI*

Bill George and John Eaves designed the *Enterprise*-B model, while the interiors were by Herman Zimmernan.

John Eaves and Herman Zimmerman were also responsible for the *Enterprise*-E for the *Star Trek: The Next Generation* movies.

Doug Drexler designed the *Enterprise* NX-01 for *Star Trek: Enterprise*, with interiors by Herman Zimmerman, who was one of *Star Trek*'s most prolific designers.

ABOUT THE AUTHORS

Between them **Ben Robinson** and **Marcus Riley** have spent many years writing about and researching *Star Trek*. They were editors and writers on the UK's phenomenally successful *STAR TREK FACT FILES* before moving on to edit (and write) *STAR TREK: THE MAGAZINE*, which ran from 1999 to 2003 in the US. Along the way they've interviewed almost everyone involved in all the *Star Trek* TV series and movies, and written about every aspect of the *Star Trek* universe from the original *Starship Enterprise* to the most obscure alien vessel.

Michael Okuda is best known for his futuristic control panels and other graphic designs for the *Enterprise* in several of the *Star Trek* films and series. He was a producer for visual effects on the remastered original *Star Trek* series and served as a technical consultant to the *Trek* writing staff. Michael lives in Los Angeles, along with his wife Denise, with whom he has co-authored several Star Trek books.

ILLUSTRATION CREDITS

Robert Bonchune 2-3, 8-11, 16-19, 40-43, 66-69, 82-85, 94-97, 106-111, 146-149, 156-159 **Richard Chasemore** 25, 45, 46-49, 52-55, 58-59, 116-117, 121, 126, 128, 130-131 **John Lawson** 13, 29, 32-33, 36-37, 51, 56-57, 62-63 (based on original art by David Kimble), 72-73, 78-79, 90-91, 98, 100-101, 135, 142-143 (based on original art courtesy CBS Studios Inc.), 155 **Haynes Publishing** 21, 24, 31, 44, 70, 86-87, 99, 114, 136-137 (based on original art by Rick Sternbach), 150 **CBS Studios Inc.** 20, 128

All photographs courtesy CBS Studios Inc.